ESP 实用视听说系列教材

【总主编 赵团结】

土木工程
英语

主　编　田妍妍

副主编　王长江　金念念

审　校　常　何　何保兴

Civil
Engineering
English

首都经济贸易大学出版社

Capital University of Economics and Business Press

·北 京·

图书在版编目(CIP)数据

土木工程英语 / 田妍妍主编. -- 北京：首都经济贸易大学出版社,2020.2

ISBN 978-7-5638-3063-3

Ⅰ.①土…　Ⅱ.①田…　Ⅲ.①土木工程—英语—教材
Ⅳ.①TU

中国版本图书馆 CIP 数据核字(2020)第 013113 号

土木工程英语
主　编　田妍妍
副主编　王长江　金念念
Tumu Gongcheng Yingyu

责任编辑　陈　侃

封面设计　**风得信·阿东**
　　　　　FondesyDesign

出版发行　首都经济贸易大学出版社

地　　址　北京市朝阳区红庙 (邮编 100026)

电　　话　(010)65976483　65065761　65071505(传真)

网　　址　http://www.sjmcb.com

E - mail　publish@ cueb.edu.cn

经　　销　全国新华书店

照　　排　北京砚祥志远激光照排技术有限公司

印　　刷　人民日报印刷厂

开　　本　787 毫米×1092 毫米　1/16

字　　数　326 千字

印　　张　13

版　　次　2020 年 2 月第 1 版　2022 年 7 月第 2 次印刷

书　　号　ISBN 978-7-5638-3063-3

定　　价　38.00 元

序 言
PREFACE

　　改革开放以后，为加强对外交往，英语被国家教育部门确定为大学必修课程。从此，万千中国学子投入大量的时间与精力去学习英语。这种长时间的重视与投入，极大地提高了国民的英语水平，密切了中外交往，加快了中国现代化进程，从而有力地促进了中国特色社会主义伟大事业的发展。经历了逾40年改革开放风雨洗礼的中国，正逐步走上世界舞台的中央。越来越有中国特色社会主义道路自信、理论自信、制度自信、文化自信的中国人民开始重新审视这种不分专业、齐头并进的全民大学英语教育，在客观、理性的探讨中已逐步形成共识，那就是：大学英语教育必须改革。可是，该如何改革呢？

　　回望过去是为了规划未来！中华儿女学习英语的热情在学习西方先进科技、开放图强的怒吼中被点燃，在追赶西方、复兴文化的脚步声中日渐高涨。时至今日，我们的英语学习怎能忘却初心、迷失方向，甚或不自觉地成为自毁文化长城的西方特洛伊木马？我们的大学英语教育改革必须把正航向。

　　知易行难！"雄关漫道真如铁，而今迈步从头越！"武昌工学院是以培养具有创新精神的应用型人才为己任的工科院校，必须以舍我其谁的气魄，在大学英语教学改革中勇当开路先锋。为此，对武昌工学院的大学英语教学改革，我提出了落实"两个一千"（1 000个专业词汇，1 000个专业句子）的具体目标。"功崇惟志，业广惟勤"，武昌工学院全体英语教师，在赵团结同志的带领下，志存高远，攻坚克难，积极推进 ESP（English for Specific Purposes）大学英语教学改革，花费大量心血，在深入调查研究的基础上，精心编写了这套"ESP 实用视听说系列教材"。"德不孤，必有邻"，我相信，这套涵盖七大专业门类的英语教材必能引起同行的关注，得到他们的爱护、响应与支持。"大厦之成，非一木之材也；大海之阔，非一流之归也"，希望本套英语教材的全体编写同志，戒骄戒躁，汇智聚力，与兄弟院校的同行携手并进，不断提高教材质量，悉心将它打造成业界精品，共同回应时代要求，开辟大学英语教学改革的新天地！

　　无尽今来古往，多少春花秋月！任何事业都是在后人赶超前人的奋斗中，不断自我更新扬弃，从而长盛不衰的。"芳林新叶催陈叶，流水前波让后波"，英语教学改革永远在路上。希望同志们不断赶超前人，持续推进英语教学改革。"合抱之木，生于毫末；九层之台，起于累土。" ESP 大学英语教学改革，已经迈出了坚实的第一步，必能善始善终，行稳致远，收前人未竟之功。

　　是为序。

武昌工学院董事长兼校长　李勇

2019 年 12 月 15 日

本书使用说明
INSTRUCTIONS

  "ESP实用视听说系列教材"是武昌工学院国际教育学院组织编写的一套教材,适用于普通高等院校本科专业大三的学生。该系列教材以学生专业学习阶段的专业词汇和专业句子为基础,以视听和口语练习为主要教学内容,强化学生的专业英语应用能力。本系列教材共7个分册,分别是:《机械工程英语》《土木工程英语》《信息工程英语》《食品化工英语》《艺术设计英语》《经济管理英语》《财务会计英语》。每个分册包括10个单元,覆盖相关专业领域内的主要专业或者方向。选用这套教材的学校,可以根据学生的专业灵活调整单元的顺序,自由选择教学的内容。

  每个单元包括六个模块,分别是:①Pre-class Activity;②Specialized Terms;③Watching and Listening;④Talking;⑤After-class Exercises;⑥Additional Reading。第一个模块是课前活动,为学生简要介绍相关领域里的名人,让学生预习,并进行自由讨论,作为课前热身。第二个模块要求学生课前预习专业词汇,教师可以在课堂上采用听写、词语接龙等形式检查学生对专业词汇的记忆效果。第三个模块是视听训练,所选视频均来自国外主流网站最新的视频资料,并根据学生的接受程度,根据由易到难的原则,编写了选择题、判断题、填空题和讨论题。第四个模块是口语训练,包括需要学生诵读的100个经典句子(50个通用句子和50个专业句子),以及供班级组织活动的两个对话样板,然后进行模仿练习和较高难度的讨论和辩论练习。第五个模块主要是结合本单元视听材料中出现的专业词汇,编写了配对、填空、翻译、写作等相关应用题型。第六个模块是课后阅读,主要内容是相关领域内的世界著名公司的简介,并采用四级考试中的长篇阅读题型,加强学生信息捕获的能力。

  每个单元大约需要6个学时。建议第一、二模块用0.5个学时,第三模块大概需要2.5个学时,第四模块需要2个学时,第五、六模块需要1个学时。为了方便师生使用本教材,我们把视频文字和练习答案,放在了出版社的网络平台上,扫描书中的二维码,即可查阅;我们还提供了相关听力视频的网址供广大师生查阅。如果无法在网上查到视频或有任何疑问和批评建议,可以联系本套丛书的总主编赵团结教授,他的联系方式如下:

手机:13986141410
座机:027-88142008
邮箱:markztj@sina.com
QQ号:3153144383
微信号:13986141410

编者

2020年1月3日

CONTENTS

Unit One   Bird's Nest

I. Pre-class Activity

Directions: *Please read the general introduction about* **Bird's Nest** *and tell something more about the* 2008 *Beijing Olympic Games to your classmates.*

Bird's Nest

The Beijing National Stadium, also known as the Bird's Nest, will be the main track and field stadium for the 2008 Summer Olympics and will be host to the opening and closing ceremonies. In 2002 Government officials engaged architects worldwide in a design competition. Pritzker Prize-winning architects Herzog & de Meuron collaborated with ArupSport and China Architecture Design & Research Group to win the competition. The stadium will seat as many as 100,000 spectators during the Olympics, but this will be reduced to 80,000 after the games. It has replaced the original intended venue of the

Guangdong Olympic Stadium. The stadium is 330 meters long by 220 meters wide, and is 69.2 meters tall. The 250,000 square meter (gross floor area) stadium is to be built with 36 km of unwrapped steel, with a combined weight of 45,000 tons. The stadium will cost up to 3.5 billion yuan (422,873,850 USD/ 325,395,593 EUR). The ground was broken in December 2003, and construction started in March 2004, but was halted by the high construction cost in August 2004.

In the new design, the roof of the stadium had been omitted from the design. Experts say that this will make the stadium safer, whilst reducing construction costs. The construction of the Olympic buildings will continue once again in the beginning of 2005.

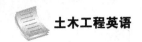

II. Specialized Terms

Directions: *Please memorize the following specialized terms before the class so that you will be able to better cope with the coming tasks.*

asphalt n. 沥青，柏油

assessment n. 评定

assign vt. 分配

assist vt. 协助

barrier n. 障碍物

basilica n. 教堂

beam n. 梁

standard specification 标准规范

burin n. 金属雕刻刀

capacity n. 容量

capture vt. 捕获，攻占

cathedral n. 大教堂

chisel v.凿，雕

church n. 教堂

circular saw 圆锯

smooth hard bedrock 光滑硬质基岩

coefficient n. (测定物质某种特性的)系数

cold chisel (凿冷金属用的)冷凿

colonnade n. 列柱，柱廊

countersink n. 锥口钻

cover n. 覆盖物，罩子

crack tip 裂纹尖端

crude adj. 粗略的，大概的

curve n. 曲线，弧线

default v.拖欠，违约

demonstrate vt. 说明，演示

design order 设计任务书

desire n. 愿望，欲望

determine v.决心，确定

durable adj. 持久的

endpoint n. 端点

energy n. 精力，[物]能量

energy release rate 能量释放率

entire adj. 全部的

envelope n. 封套，封袋

existing adj. 现存的，目前的

fatigue in metals 金属疲劳

file n. 文件，档案

firmer gouge 冰凿

float n. 抹子

folder n. 文件夹

folding ruler 折尺

fork n. 叉子，岔口

format vt. 设计，安排

fraction n. 碎片，片段

fretsaw n. 线锯，圆锯

friction n. 摩擦(力)

graphically adv.用图表表示地

greenbelt n. 绿地

grid snap 栅格捕捉

initial adj. 最初的

innovation n. 创新，改革

installation n. 装置，设备

interact vi. 相互作用

interval n. 间隔，距离

invade v.拥入，大批进入

invoke vt. 使产生

jack plane 粗刨

kinetic adj. 运动引起的

mallet n. 木槌

miter n. 斜接

monument n. 纪念碑

moulding plane 型刨

multiple adj. 数量多的

nail n. 钉
nail puller 拔钉器
navigate vi. 航海,导航
negotiation n. 谈判
outcrop n. 露出地面的岩层
outline n. 提纲;外形,轮廓,略图
overview n. 概观,概述
paintbrush n. 画刷
palace n. 宫殿
pier n. 墩,桥墩
plane n. 水准,刨子
planning proposals 计划建议书
principle of virtual work 虚功原理
probability n. 概率
proceed vi. 继续进行
processor n. 信息处理机
projectile n. 抛射物,发射体
prompt vt. (to)促使,推动,提示
prompt line 提示行

property n. 财产,地产
purpose n. 目的,用途
rabbet plane 槽刨
rasp n. 锉,锉刀
raw adj. 生的,未加工的
rockfall n. 岩崩,大量的落石
roller n. 滚筒,滚轴
roughness n. 粗糙
ruler n. 尺
sandpaper n. 砂纸
scheme n. 方案
time-dependent deformation 时相关变形
toolbar n. 工具条(栏)
toolbox n. 工具箱
tower n. 塔,塔楼
town planning 市政(规划)
tread v.踩,踏,践踏
trestle n. 架柱, 支架

III. Watching and Listening

Task One   Bird's Nest (I)

New Words

daring adj. 勇敢的
beam n. 横梁;电波
weave vt. 使迂回前进
stunning adj. 震耳欲聋的
vision n. 视力
gigantic adj. 巨大的
arena n. 竞技场

massive adj. 大量的
spectator n. 观众
unprecedented adj. 空前的
curving adj. 弯曲的
crisscross n. 十字形
intricate adj. 复杂的

视频链接及文本

Exercises

1. *Watch the video for the first time and answer the following questions.*

   1) What is the nickname of the Beijing's Olympic Stadium? _____.

      A. Bird's neck   B. Bird's head

      C. Bird's mouth   D. Bird's nest

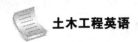

土木工程英语

2) How many tons of steel beams were wove together into the building? _____.

    A. 42 thousand tons   B. 40 thousand tons

    C. 14 thousand tons   D. 22 thousand tons

3) How long is the steel roof? _____.

    A. 1,015 feet long   B. 1,050 feet long

    C. 115 feet long   D. 150 feet long

4) How high is the structure? _____.

    A. 213 feet   B. 233 feet

    C. 230 feet   D. 330 feet

5) How many people can sit in the stadium? _____.

    A. 19 thousands   B. 191 thousands

    C. 910 thousands   D. 91 thousands

2. *Watch the video again and decide whether the following statements are true or false.*

1) Today workers must raise a gigantic mass of twisted steel to the middle of this structure. (　)

2) The stadium takes center stage in one of the biggest events of this year. (　)

3) It is the center of the biggest sporting event in Chinese history the 2008 Summer Olympics. (　)

4) In the center of Beijing, lies a daring old design. (　)

5) It is the largest stadium ever built in China. (　)

3. *Watch the video for the third time and fill in the following blanks.*

1) It's the stunning vision of international team of _____ and _____.

2) Today workers must raise a gigantic mass of _____ steel to the top of this structure.

3) It's the last piece of the _____ to complete the steel _____ of this unusual _____.

4) And it is the center of the biggest _____ event in Chinese history the 2008 _____ Olympics.

5) It's an ambitious plan that calls for giant curving _____, which crisscross in an intricate _____ of woven steel.

4. *Share your opinions with your partners on the following topic for discussion.*

    What elements should be considered in building a stadium?

Task Two　Bird's Nest (Ⅱ)

视频链接及文本

New Words

evolve v.发展,使逐步形成

fabricate v.制造

bells and whistles 附加的修饰物

seismic adj. 地震引起的

fragile adj. 脆的

withstand v.抵挡,反抗

mandate n. 授权,委托管理

membrane n. 膜,羊皮纸

4

facade n. 正面,外观　　　　　　　　　　radical adj. 激进的

ventilation n. 通风设备

Exercises

1. *Watch the video for the first time and answer the following questions.*

   1）The woman says that the Bird's Nest whether its twists and turns and complications is certainly one of the ones that's way out on the edge,_____.

   A. first of its kind in China, first of its kind in the world

   B. first of its kind in Beijing, first of its kind in China

   C. first of its kind in China, first of its kind in the Asia

   D. first of its kind in China, second of its kind in the world

   2）What must an Olympic stadium have? _____.

   A. Giant ground, shops and great views of the field

   B. Giant screens, shops and great views of the field

   C. Giant ground, windows and great views of the field

   D. Giant screens, windows and great views of the field

   3）What have officials mandated this Olympics to be? _____.

   A. Green Olympics　　　　　　　　B. Blue Olympics

   C. Yellow Olympics　　　　　　　　D. White Olympics

   4）What is the steel roof skinned with? _____

   A. Hi-tech cloth　　　　　　　　　B. Hi-tech membranes

   C. Hi-tech silk　　　　　　　　　　D. Hi-tech feather

   5）Why does the stadium let the sunlight in? _____

   A. To consume energy　　　　　　　B. To solve energy

   C. To conserve energy　　　　　　　D. To attract energy

2. *Watch the video again and decide whether the following statements are true or false.*

   1）Maybe the stadium is one of the biggest wood structures in the world. (　)

   2）I will not see any other projects easier than this Bird's Nest. (　)

   3）The Bird's Nest must have all the bells and whistles of any Olympic stadium. (　)

   4）The stadium's facade is enclosed. (　)

   5）All well up against the movable deadline of the 2008 Summer Olympics. (　)

3. *Watch the video for the third time and fill in the following blanks.*

   1）No _____ like this one has ever been _____ before in the world.

   2）It's quite clear that it was going to be a major challenge to _____ it, and to fabricate it, and to _____ it.

   3）Whether its _____ and turns and _____ is certainly one of the ones that's way out on the edge.

   4）The _____ design must be strong enough to withstand dangerous earthquakes and it

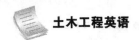

must be _____.

5) State-of-the-art systems _____ energy and water and _____ heat.

4. *Share your opinions with your partners on the following topic for discussion.*

What does your ideal stadium look like?

IV. Talking

Task One   Classical Sentences

Directions：*In this section, some popular sentences are supplied for you to read and to memorize. Then, you are required to simulate and produce your own sentences with reference to the structure.*

General Sentences

1. Where do you live?

你住在哪里?

2. I live in Washington Street.

我住在华盛顿街。

3. I'm Mr. Smith's neighbor.

我是史密斯先生的邻居。

4. You live in the city, don't you?

你住在这个城市,对吗?

5. I live outside the city.

我住在城外。

6. How long have you lived here?

你在这住了多久了?

7. I've lived here for five years.

我在这住了5年了。

8. Where did you grow up?

你在哪长大?

9. I grew up right here in this neighborhood.

我就在这附近长大。

10. My friend spent his childhood in California.

我朋友是在加利福尼亚度过他的童年的。

11. He lived in California until he was seventeen.

他17岁以前都在加利福尼亚住。

12. There have been a lot of changes here in the last 20 years.

在过去20年间,这里发生了很多的变化。

13. There used to be a grocery store on the corner.
以前拐角处有一个杂货店。

14. All of those houses have been built in the last ten years.
那些房子都是在最近十年里建成的。

15. They're building a new house up the street.
这个街上正在建一个新房子。

16. If you buy that house, will you spend the rest of your life there?
如果你买了那栋房子,是不是打算就在那里住到老?

17. Are your neighbors friendly?
你的邻居们友好吗?

18. We all know each other pretty well.
我们都很了解对方。

19. Who bought that new house down the street?
大街那头的那栋新房子谁买下了?

20. An old man rented the big white house.
一个老人租了那栋白色大房子。

21. We're looking for a house to rent for the summer.
我们在找一栋房子夏天租住。

22. Are you trying to find a furnished house?
你是不是在找一栋带装修的房子?

23. That house is for sale. It has central heating. It's a bargain.
那栋房子在出售。房子是集中供暖,价格很合理。

24. This is an interesting floor plan. Please show me the basement.
这是一个不错的平面图。麻烦带我到地下室去看看。

25. The roof has leaks in it and the front steps need to be fixed.
屋顶漏水,前面台阶也需要修理。

26. We've got to get a bed and a dresser for the bedroom.
我们得在卧室弄张床和一个梳妆台。

27. They've already turned on the electricity. The house is ready.
房子已经通电,可以入住了。

28. I'm worried about the appearance of the floor.
我对地板的外观甚感担忧。

29. What kind of furniture do you have? Is it traditional?
你这里有什么样式的家具? 是传统型的吗?

30. We have drapes for the living room, but we need kitchen curtains.
客厅的窗帘我们有了,但我们需要厨房的窗帘。

31. The house needs painting. It's in bad condition.
这房子得粉刷了,情况很糟。

32. In my opinion, the house isn't worth the price they're asking.
依我看,这间房子根本不值他们要的价钱。

33. Will you please measure this window to see how wide it is?
请你测量一下这个窗户,看看它有多宽?

34. This material feels soft.
这种材料摸上去很软。

35. —Can you tell me where Peach Street is?
—Two blocks straight ahead.
—你能告诉我皮彻大街在哪儿吗?
— 一直朝前走,过两个街区就到了。

36. Should I go this way, or that way?
我要走这条路还是那条?

37. Go that way for two blocks and then turn left.
走那条路,穿过两个街区后向左拐。

38. How far is it to the university?
到大学还有多远?

39. The school is just around the corner. It's a long way from here.
学校就在拐角处。从这儿走,还有很长的一段路。

40. Are you married?
你结婚了吗?

41. No, I'm not married. I'm still single.
没有,我还是单身。

42. Your niece is engaged, isn't she?
你的侄女订婚了,是吗?

43. When is your grandparents' wedding anniversary?
你祖父母的结婚纪念日是什么时候?

44. How long have they been married?
他们结婚多久了?

45. They've been married for three decades.
他们结婚30年了。

46. We're trying to plan our future.
我们在努力计划我们的未来。

47. I've definitely decided to go to California.
我已经决定去加利福尼亚了。

48. Who are you writing to?
你在给谁写信呢?

49. I'm writing to a friend of mine in South American.
我在给我南美的一位朋友写信。

50. —How long has it been since you've heard from your uncle?

—I feel guilty because I haven't written to him lately.

—从你收到你叔叔的信到现在多久了?

—我总觉得很内疚,因为我最近没有给他写信。

Specialized Sentences

1. It is very fine today and the weather is suitable for our work.

今天天气很好,适合我们工作。

2. We cannot continue the outdoor work, because it is raining now.

因为现在下雨,我们不能继续在室外工作。

3. It is going to storm tomorrow and some measures must be taken to prevent wind.

明天将起风暴,必须采取一些防风措施。

4. The lifting work on site will be compelled to stop, owing to the strong wind.

由于强风,现场起重吊装工作将被迫停止。

5. This road leads to the factory.

这条路通到工厂。

6. There is an article of equipment in front of the building.

在建筑物前面有一台设备。

7. Our construction site is north of Yanbu, near the Red Sea.

我们的工地在延布以北,靠近红海。

8. Today we shall discuss the question of hydraulic test.

今天我们将讨论水压试验问题。

9. This problem must be reported to the higher level, they have the final say to make decisions.

这个问题应该呈报上级,由他们做最后决定。

10. Your side will be held responsible for all the consequences arising from there.

由此产生的一切后果由你方负责。

11. Let us draft a resolution about it.

让我们为此起草一项决议。

12. A project execution is usually divided into some elementary phases, such as, engineering planning, procurement and transportation, field construction and pro-commissioning.

一个工程项目的实施通常可分为几个基本阶段:工程设计、采购和运输、现场施工和调试。

13. We are building a polypropylene plant with an annual capacity of 400,000 metric tons.

我们正在建设一座年产 40 万吨的聚丙烯工厂。

14. The contract number of this project is EPC-LS-ORYANSAB-010-IK.

这个项目的合同号是 EPC-LS-ORYANSAB-010-IK。

15. The client is SABIC.

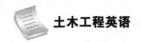

委托方是沙特基础工业公司。

16. The buyer is China National Technical Import Corporation(CNTIC).

买方是中国技术进口总公司(简称 CNTIC)。

17. Sinopec contracts for domestic and overseas chemical projects.

中国石化集团承包国内和海外的化工工程。

18. We can contract to build turn-key plant, undertake single items of projects as a subcontractor or provide labor services.

我们可以承建成套工厂,可以分包单项工程或提供劳务。

19. The Dow takes part in this project as a patent licensor.

美国陶氏公司作为专利授权人参加此项工程。

20. There are two units in the contract plant.

在合同工厂内有两个车间。

21. The lifting work on site will be compelled to stop, owing to a dense fog.

由于浓雾,现场起重吊装工作将被迫停止。

22. Some important equipment behind the water tower is being repaired.

水塔后面的一些重要设备正在维修。

23. There is a digger under the pipe rack.

在管廊下面有一台挖掘机。

24. The engineer is testing the equipment on the floor.

工程师正在检验地面上的那台设备。

25. Who can help me move the equipment inside the steel structure out?

谁能帮我把钢结构里面的那台设备移出来?

26. No one knows where the equipment in the workshop has gone.

没有人知道车间内那台设备去哪里了。

27. Today we shall discuss the question of pneumatic test.

今天我们将讨论气压试验问题。

28. We are building a linear low density polyethylene plant with an annual capacity of 400,000 metric tons.

我们正在建设一座年产 40 万吨的线性低密度聚乙烯工厂。

29. The two parties are talking about the possibility of building a product handling plant with an annual capacity of 400,000 metric tons.

双方正在讨论,是否有可能建设一座年产 40 万吨的产品包装工厂。

30. There are two installations within the battery limits.

在电池负载内有两个装置。

31. The project is certain to be a success.

这个项目一定会取得成功。

32. The effective date of this contract will begin from July the 20, 2005.

这个合同的有效期将从 2005 年 7 月 20 日开始。

33. On our most projects, Critical Path Method(CPM) is used for scheduling.

我们的大多数工程项目中都采用"统筹法"(即关键路线方法,简称 CPM)安排计划。

34. Civil work will begin in June this year and complete on the end of December next year.

土建工程将自今年 6 月开始至明年 12 月底完工。

35. This contract plant will start-up on January 1st, 2008.

这座合同工厂将于 2008 年 1 月 1 日开工。

36. This contract plant will put in commissioning on January first 2012.

这座合同工厂将于 2012 年 1 月 1 日投产。

37. The date of acceptance of this plant will be December 31,2007.

这座工厂的交工验收期将在 2007 年 12 月 31 日。

38. The plant is scheduled to be completed around 2008.

工厂计划于 2008 年前后建成。

39. This is similar to the preceding feature but applies to tool length and diameter.

这是类似上述的功能,但适用于刀具长度和直径。

40. These actual dimensions may differ from those originally programmed.

最初的方案与这些实际尺寸可能会有所不同。

41. Compensations are then automatically made in the computed tool path.

然后通过自动计算刀具路径提供代偿。

42. This measured value is then used to correct the programmed tool path.

然后用此测量值修正方案中的刀具路径。

43. One of the biggest problems when a machine failure occurs is often in diagnosing the reason for the breakdown.

一台机器发生故障时最大的难题经常就是如何诊断故障原因。

44. By monitoring and analyzing its own operation, the system can determine and communicate the reason for the failure.

系统可以通过自带的运行监测和分析功能,判断和通报失败的原因。

45. Another use of the diagnostics capability is to help the repair crew determine the reason for a breakdown of the machine tool.

诊断能力的另一个用途是帮助维修人员确定机床出问题的原因。

46. In older-style controls, the cutter dimensions had to be set very precisely in order to agree with the tool path defined in the part program.

在老式的控制中,刀具尺寸设置非常精确,以符合在零件程序中对刀具路径的界定。

47. Setting up the machine tool for a certain job involves installing and aligning the fixture on the machine tool table.

要完成一定的工作,需要安装和校准安装在机床的工作台上的夹具。

48. This must be accomplished so that the machine axes are aligned with the workpart.

该项必须完成,以使机器轴与工件对齐。

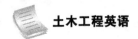

Task Two　Sample Dialogue

Directions: *In this section, you are going to read several times the following sample dialogue about the relevant topic. Please pay special attention to five C's (culture, context, coherence, cohesion and critique) in the sample dialogue and get ready for a smooth communication in the coming task.*

(Mr. Zhang, assistant of the project manager, is making telephone calls, trying to rent a house for the accommodations of the crew.)

Zhang: Hello, may I speak to Mr. Neal, please?

Neal: Speaking. What can I do for you, sir?

Zhang: I'm calling about your advertisement in *People's Pulse*. Have you got a house for rent?

Neal: Well, I did have a house for rent. But it's been let out.

Zhang: Sorry to hear that. By the way, do you know anyone who has a house for rent nearby?

Neal: Let me see. My friend Sam told me a few weeks ago that he wanted to rent his house. I think it's a big house. You could check with him.

Zhang: That's just what we're looking for. What's his telephone number, please?

Neal: 2237.

Zhang: Could I have your friend's name again?

Neal: Sam, Sam Carter.

Zhang: Thanks a lot. Bye.

*　　*　　*　　*　　*　　*　　*　　*

Zhang: Good morning. I'd like to speak to Mr. Carter, please.

Man: Sorry, there's no one here by that name.

Zhang: What's your telephone number then?

Man: 2337.

Zhang: I'm very sorry. I must have dialed the wrong number.

*　　*　　*　　*　　*　　*　　*　　*

Zhang: Hello, is Mr. Carter there?

Woman: Just a minute. He's coming.

Carter: Hello. Carter speaking. Who is calling, please?

Zhang: My name's Zhang. I have been told that you have a house for rent. Is it still available?

Carter: Yes. It's a two-story building with six bed rooms on the first floor and two on the ground floor. It has two bathrooms, one on each floor. There's a good kitchen where you could do cooking. We also have a cable TV and a telephone that you can use in the house.

Zhang: Perhaps it's just what we need. What's the monthly rent?

Carter: How long will the lease be?

Zhang: About four months.

Carter: Then two thousand dollars. And the rent is due at the end of each month. You can use all the utilities, but you have to pay all the bills including the telephone service, cable TV, water, gas and electricity. By the way, you also have to pay a security deposit of one thousand dollars.

Zhang: Is there any space for office work in the house?

Carter: There is a big sitting room on the ground floor. I think it's where you can do your office work.

Zhang: One more thing, are we allowed to install a fax machine in the house if we take it? Probably we will need one.

Carter: Yes, if you like. But you have to pay all the costs for it.

Zhang: Good. Would it be possible for us to go there and have a look tomorrow?

Carter: Any time before four o'clock in the afternoon. You can come to Okis Restaurant. I'm running here in Benque, Cayo District. It's on George Street. Benque is a small town. Everyone here knows Okis. I can wait for you here.

Zhang: OK. We will go there tomorrow, around ten o'clock in the morning. Will that be all right with you?

Carter: OK. See you tomorrow, sir.

Zhang: Thank you. Mr. Carter, See you tomorrow morning.

Task Three   Simulation and Reproduction

Directions: *The class will be divided into three major groups, each of which will be assigned a topic. In each group, some students may be the teacher, while others may be students. In the process of discussion, please observe the principles of cooperation, politeness and choice of words. One of the groups will be chosen to demonstrate the discussion to the class.*

1) Stadium in your school or hometown.

2) A funny story related to stadium in your life.

3) The importance of learning engineering.

Task Four   Discussion and Debate

Directions: *The class will be divided into two groups. Please choose your stand in regard to the following controversy and support your opinions with scientific evidences. Please refer to the specialized terms and classical sentences in the previous parts of this unit.*

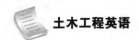

In this unit, we have already talked about the Bird's Nest. However, there is another famous building in 2008 Olympic Games—Water Cube. Do some research on these two fabulous stadiums. Which one do you prefer? Why?

V. After-class Exercises

1. *Match the English words in Column A with the Chinese meaning in Column B.*

A	B
1) planning proposals	A) 原始资料
2) standards and codes	B) 标准规范
3) basic data for design	C) 可行性研究
4) engineering geological data	D) 设计进度
5) original data	E) 工程地质资料
6) schedule of design	F) 设计基础资料
7) summary of discussion	G) 计划建议书
8) negotiation	H) 初步设计
9) feasibility study	I) 会谈纪要
10) preliminary design	J) 谈判

2. *Fill in the following blanks with the words or phrases in the word bank. Change the forms if it's necessary.*

correct	schedule	certain	report	measure
limit	commission	responsible	article	compel

1) It is going to storm tomorrow and some measures must be taken to _____ wind.

2) The lifting work on site will be _____ to stop, owing to the strong wind.

3) There is an _____ of equipment in front of the building.

4) This problem must be _____ to the higher level, because they have the final say to make decisions.

5) Your side will be held _____ for all the consequences arising there from.

6) The project is _____ to be a success.

7) This contract plant will put in _____ on January first, 2012.

8) The plant is _____ to be completed around 2008.

9) There are two installations within the battery _____.

10) This measured value is then used to _____ the programmed tool path.

3. *Translate the following sentences into English.*

1) 今天我们将讨论水压试验问题。

_____

_____

2）土建工程将自今年6月开始至明年12月底完工。

_____

_____

3）最初方案与这些实际尺寸可能会有所不同。

_____

_____

4）该项必须完成，以使机器轴与工件对齐。

_____

_____

5）这是类似上述的功能，但适用于刀具长度和直径。

_____

_____

4. *Please write an essay of about* 120 *words on the topic：**Becoming an engineer.** Some specific examples will be highly appreciated and watch out the spelling of some specialized terms you have learnt in this unit.*

_____

_____

_____

_____

_____

_____

_____

_____

_____

VI. Additional Reading

Bird's Nest

［A］Located in the Olympic Green, the stadium cost US $428 million. The design was awarded to a submission(呈递) from the Swiss architecture firm Herzog & de Meuron in April 2003, after a bidding process that included 13 final submissions. The design, which originated from the study of Chinese ceramics(制陶工艺), implemented steel beams in order to hide supports for the retractable(可开合的) roof; giving the stadium the appearance of a "Bird's

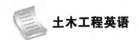

nest". Leading Chinese artist Ai Weiwei was the artistic consultant on the project.

[B]The retractable roof was later removed from the design after inspiring the stadium's most recognizable aspect. Ground was broken on 24 December 2003 and the stadium officially opened on 28 June 2008. A shopping mall and a hotel are planned to be constructed to increase use of the stadium, which has had trouble attracting events, football and otherwise, after the Olympics.

[C]In 2001, before Beijing had been awarded, the city held a bidding process to select the best arena design. Multiple requirements including the ability for post-Olympics use, a retractable roof and low maintenance costs, were required of each design. The entry list was narrowed to thirteen final designs.

[D]Of the final thirteen, Li Xinggang of China Architecture Design and Research Group (CADG) said after he placed the model of the "nest" proposal at the exhibition hall and saw the rival entries he thought to himself, "We will win this." The model was approved as the top design by a professional panel and later exhibited for the public. Once again, it was selected as the top design. The "nest scheme" design became official in April 2003.

[E]Beijing National Stadium (BNS) was a joint venture among architects Jacques Herzog and Pierre de Meuron of Herzog & de Meuron, project architect Stefan Marbach, artist Ai Weiwei and CADG which was led by the chief architect Li Xinggang. During their first meeting in 2003, at Basel, the group decided to do something unlike Herzog and de Meuron had traditionally designed. "China wanted to have something new for this very important stadium", Li stated.

[F]In an effort to design a stadium that was "porous(多孔的)" while also being "a collective building, a public vessel", the team studied Chinese ceramics. This line of thought brought the team to the "nest scheme". The stadium consists of two independent structures, standing 50 feet apart: a red concrete seating bowl and the outer steel frame around it.

[G]In an attempt to hide steel supports for the retractable roof, required in the bidding process, the team developed the "random-looking additional steel" to blend the supports into the rest of the stadium. Twenty-four trussed(架子) columns encase the inner bowl, each one weighing 1,000 tons. Despite the random appearance of the Stadium, each half is nearly symmetrical(对称的). After a collapse of a roof at the Charles de Gaulle Airport, Beijing reviewed all major projects.

[H]It was decided to eliminate the retractable roof, the original inspiration for the "nest" design, as well as 9,000 seats from the design. The removal of the elements helped to bring the project under the reduced construction budget of $290 million, from an original $500 million. With the removal of the retractable roof, the building was lightened, which helped it stand up to seismic(地震的) activity; however, the upper section of the roof was altered to protect fans from weather.

[I]Enerpac was granted the contract to perform the stage lifting and lowering of the stadium roof as part of the construction process. China National Electric Engineering Co. Ltd. (CNEEC) and China National Mechanical Engineering Company lifted and welded (焊接) the steel structure. Due to the stadium's outward appearance, it was nicknamed "Bird's Nest". The phrase was first used by Herzog & de Meuron, though the pair still believes "there should be many ways of perceiving a building". The use is a compliment Li explained, "In China, a bird's nest is very expensive, which is something you eat on special occasions."

[J] Construction of the stadium proceeded in several distinct phases, the first phase involving the construction of a concrete supporting structure upon the concrete foundations laid for the construction site. This was followed by the phased installation of the curved steel frame surrounding the stadium which is largely self-supporting. This phased installation involved the interconnection of sections of the curved steel frame which were constructed in Shanghai and transported to Beijing for assembly and welding. The entire structure of interconnected sections was welded together as the primary means of interconnection used to assemble the entire surrounding nest structure. Upon removal of the supporting columns used for the purpose of expediting the assembly of the interconnecting sections, the completed nest structure as a whole settled approximately 27 cm to attain full stability before the interior design and construction of the stadium could be installed and completed.

[K] Ground was broken, at the Olympic Green, for Beijing National Stadium on 24 December 2003. At its height, 17,000 construction workers worked on the stadium. Portraits of 143 migrant workers at the construction site were featured in the book *Workers* (*Gong Ren*) by artist Helen Couchman. All 121,000 tons of steel were made in China. On 14 May 2008 the grass field of 7,811 square meters was laid in 24 hours. The field is a modular (组合式的) turf (草皮) system by GreenTech ITM. Beijing National Stadium officially opened at a ceremony on 28 June 2008.

1. *Read the passage quickly by using the skills of skimming and scanning. And choose the best letter standing for each paragraph above in response to the following sentences.*

   1) Leading Chinese artist Ai Weiwei was the artistic consultant on the project.

   2) The removal of the elements helped to bring the project under the reduced construction budget of \$290 million, from an original \$500 million.

   3) Enerpac was granted the contract to perform the stage lifting and lowering of the stadium roof as part of the construction process.

   4) The entire structure of interconnected sections was welded together as the primary means of interconnection used to assemble the entire surrounding nest structure.

   5) On 14 May 2008, the grass field of 7,811 square meters was laid in 24 hours.

   6) In 2001, before Beijing had been awarded the games, the city held a bidding process to select the best arena design.

   7) Once again, it was selected as the top design. The "nest scheme" design became

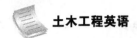

official in April 2003.

8) During their first meeting in 2003, at Basel, the group decided to do something unlike Herzog and de Meuron had traditionally designed.

9) Ground was broken on 24 December 2003 and the stadium officially opened on 28 June 2008.

10) The stadium consists of two independent structures, standing 50 feet apart: a red concrete seating bowl and the outer steel frame around it.

1) _____ 2) _____ 3) _____ 4) _____ 5) _____

6) _____ 7) _____ 8) _____ 9) _____ 10) _____

2. *In this part, the students are required to make an oral presentation on either of the following topics.*

1) The secrets of Bird's Nest's design.

2) The lessons from Bird's Nest's construction.

习题答案

Unit Two   Bidding Documents

I. Pre-class Activity

Directions: *Please read the general introduction about* **William LeMessurier** *and tell something more about the great structural engineer to your classmates.*

William LeMessurier

William LeMessurier (1926~2007) was a prominent American structural engineer. Born in Pontiac, Michigan, LeMessurier graduated with a BA from Harvard, went to Harvard Graduate School of Design and then earned a master's degree from Massachusetts Institute of Technology in 1953. He was the founder and the chairman of LeMessurier Consultants. He was awarded the AIA Allied Professions Medal in 1968, elected to the National Academy of Engineering in 1978, elected an honorary member of the American Institute of Architects in 1988 and elected an honorary member of the American Society of Civil Engineers (ASCE) in 1989. In 2004, he was elevated to National Honor Member of Chi Epsilon, the national civil engineering honor society.

While responsible for the structural engineering on a large number of prominent buildings, including Boston City Hall, the Federal Reserve Bank of Boston, the Singapore Treasury Building and Dallas Main Center, LeMessurier is perhaps best known for a structural controversy. As the result of the questions of a student (Diane Hartley), LeMessurier re-assessed his calculations on the Citicorp headquarters tower in New York City in 1978, after the building had already been finished and found that the building was more vulnerable than originally thought (in part due to cost-saving changes made to the original plan by the contractor). This triggered a hurried, clandestine retrofit which was described in a celebrated article in the New Yorker. The article, titled "The Fifty-Nine-Story Crisis", is now used as an ethical case-study. LeMessurier died in Casco, Maine, on June 14, 2007 as a result of complications after surgery he underwent on June 1 after a fall the day before.

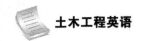

II. Specialized Terms

Directions: *Please memorize the following specialized terms before the class so that you will be able to better cope with the coming tasks.*

abandoned hole 废孔

adit n. 坑道,出入口

advance depth 进尺深度

aggravate v. 使严重

average hole depth 平均孔深

as-built drawing 竣工图

basic data for design 设计基础资料

basic design 基础设计

bench excavation 台阶式挖掘

bench n. (木制或石制)长凳

blank adj. 无图画(或标记)的

blasting v. 爆破

block of flats 公寓楼

bounce vt. (使)弹起,(使)反弹

bounce height 弹跳高度

commerce n. 商务,商业

column n. 柱

compass saw 钢丝锯

compression n. 压缩,压挤

concrete n. 混凝土

construct vt. 建设,建筑,修建

convenience n. 便利设施[复]

cushion hole 缓冲孔

cutoff trench 截水槽

detail design 细节设计

deviation n. 偏差

dialog n. (=dialogue)会话,对话

distribution n. 分配

drawknife n. 刮刀

drill bit 钻(头,尖)

drift n. 掏槽

drilling n. 钻孔

driving (progress) rate 进尺率

ductile material 韧性材料

engineering geological data 工程地质资料

expatriate n. 外籍职员

enquiry drawing 询价图

feasibility study 可行性研究

instrumentalist n. 工具

information drawing 信息制图

labor n. 劳务

mandatory adj. 强制的

manner n. 方式,方法

manual n. 使用手册

margin n. 边缘

marking gauge 测量仪器(或仪表)

masonry n. 石工技术,石屋

objection n. 反对

obligation n. 责任,义务

perimeter hole 周边孔

preoccupation n. 当务之急;抢先占据

requalification n. 再鉴定

original data 原始资料

preliminary design 初步设计

preside n. 主持,主管

process flowchart 工艺流程图

prominent adj. 突起的,凸出的

recommend vt. 推荐,介绍

redirect v.使改方向

remedial adj. 补救的

reorient 重定……的方向

repeat v.重做

restitution n. 归还,赔偿

rigidly adv.坚硬地,严格地

schedule of design 设计进度

screen capture 截图,截屏

screw tap 螺丝攻

screw n. 螺丝钉,螺丝

screwdriver n. 螺丝起子,改锥

scriber n. 描绘标记的用具, 画线器, 近线尺

scroll saw 钢丝锯

seeder n. 播种机, 去核器

spatula n. 抹刀

specified adj. 指定的

speedy adj. 快速的

spreadsheet n. 电子数据表, 电子制表软件, 空白表格程序

square n. 正方形,直角尺

standard deviation 标准方差,标准差,标准偏差

statistical adj. 统计的, 统计学的, 统计上的

stepladder n. 折梯,梯子

strain energy 应变能

summary of discussion 会谈纪要

tabular adj. 制成表的,平坦的

tack n. 平头钉,大头钉,图钉

talus n. 斜面,距骨,碎石堆

tangential adj. 切线的, 相切的, 正接的

tape puller 卷尺

ten-storey office block 十层办公大楼

timber n. 木材,木料

variation n.(数量、水平等的)变化,变更

various adj. 不同的,多方面的

vegetation n. 植被

velocity n. 速度, 速率, 迅速

construction drawing 施工图

III. Watching and Listening

Task One   The Life of Construction Project Management

New Words

project n. 项目,工程

technical adj. 技术(性)的

multitasking n. 多(重)任务处理

assemble v.集合

consensus n. 一致同意

agenda n. 议事日程

execute v.履行

collaborative adj. 合作的, 协作的

recipe n. 秘诀

视频链接及文本

Exercises

1. *Watch the video for the first time and choose the best answers to the following questions.*

1) A typical day for a project manager can start _____.

　　A. from making phone calls　　　　B. from having some meetings

　　C. in a variety of places　　　　　D. from keeping up on email

2) To be a project manager needs the following feature except _____.

　　A. cooperating spirit　　　　　　B. technical skills

　　C. strong communication　　　　　D a passion for multitasking

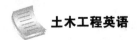

3) To help people work together towards a common goal involves the following except _____.

    A. listening   B. sharing idea

    C. creating a sense of accomplishment   D. setting clear goals and objectives

4) According to Jimmy Ali technology will play a big part in the construction industry in the following parts except _____.

    A. time saving   B. money saving

    C. offering opportunities for additional careers   D. labor savings

5) The secret of success in the teamwork is _____.

    A. everybody does his own work individually

    B. respecting each other's ideas and opinions

    C. being collaborative

    D. to make each other successful

2. *Watch the video again and decide whether the following statements are true or false.*

1) JH Finned's project manager's day started at the conference. (　　)

2) JH Finned, an orphan project manager, have no time for some paperwork. (　　)

3) Being a project manager requires strong communication, technical skills and enthusiasm for multitasking. (　　)

4) Technology plays a big role in today's construction industry. (　　)

5) Using computers does not mean saving time and money on all projects. (　　)

3. *Watch the video for the third time and fill in the following blanks.*

A typical day for a project _____ can start in a variety of places. For me it usually starts in the office where I make phone calls and keep up on _____. I will usually have the day laid out with some _____, some in the office and some on _____ and also some time for some _____ work. To be a project manager, take strong communication, _____ skills and a passion for _____. Another part of my day was important is to pull together meetings. Today we've _____ the team here that's going to work on a project it chanced them to share the _____ and pull together before we start the _____.

4. *Share your opinions with your partners on the following topic for discussion.*

1) How do you feel the day of being a project manager? Please summarize the features of project management.

2) Can you use a few lines to list your understanding about tenders? Please use an example to clarify your thoughts.

Task Two　Online Professional Certificates in Construction

New Words

supervisor n. 监督者   brand-new adj. 全新的

administrative adj. 管理的   overall adj. 综合的

contractor n. 承包人   certificate n. 证明书

paycheck n. 薪水,工资支票   consultant n. (受人咨询的)顾问

Exercises

1. *Watch the video for the first time and choose the best answers to the following questions.*

   1) What kinds of person will come to watch the online class? _____.

      A. Supervisors

      B. Brand-new estimators

      C. Folks who want to jump into the industry

      D. Different kinds of people

   2) Why do people suggest Wendy take these courses as a transition from administrative work more into project management? _____.

      A. Because the program is offered online

      B. Because they can access it on their lunch break

      C. Because these classes are truly for anyone and everyone

      D. Because these classes are very popular

   3) These classes are best choice because of the following reasons except _____.

      A. the program is offered online

      B. it is convenient for everyone

      C. the instructors are industry experts

      D. people can log in at night after the kids are in bed

   4) Children benefit from these courses _____.

      A. in their bottom line

      B. in their paycheck

      C. in their skills

      D. to raises their confidence

   5) The theme of this passage is _____.

      A. the more education you receive the better you are

      B. online classes are truly for anyone and everyone

      C. students can benefit from these classes a lot

      D. construction industry should learn more about online classes

2. *Watch the video again and decide whether the following statements are true or false.*

   1) Wendy Fitzgerald is in the mechanical industry. (   )

   2) It is very inconvenient for me to take these courses at home online.(   )

   3) The instructors are experienced industry experts. (   )

   4) The instructors have a lot of resources that are beyond the theory. (   )

   5) The students have failed to develop a confidence in the skills. (   )

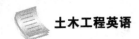

3. *Watch the video for the third time and fill in the following blanks.*

The folks who come to us, they're _____, they're brand-new _____, they're folks who want to jump into the industry because they know about the _____ that we have. These classes are truly for anyone and everyone. My name is Wendy Fitzgerald and I'm in the _____ industry. I was working for, a general _____, at the time and they suggested that I take these courses as a _____ from administrative work more into project _____. Because the program is offered _____, students love it because they can _____ it on their lunch break, first thing in the morning before the _____ gets in.

4. *Share your opinions with your partners on the following topic for discussion.*

  1) Do you know how to prepare bidding documents?

  2) Can you design a tender with the aid of computer?

IV. Talking

Task One   Classical Sentences

Directions: *In this section, some popular sentences are supplied for you to read and to memorize. Then, you are required to simulate and produce your own sentences with reference to the structure.*

General Sentences

  1. What's your nationality? Are you Chinese?

    你是哪国人? 你是中国人吗?

  2. What part of the world do you come from?

    你来自哪里?

  3. I was born in Spain, but I'm a citizen of France.

    我出生在西班牙,但我是法国公民。

  4. Do you know what the population of Japan is?

    你知道日本有多少人口吗?

  5. What's the area of the Congo in square miles?

    刚果的面积是多少平方英里?

  6. Who is the governor of this state?

    谁是该州的州长?

  7. According to the latest census, our population has increased.

    根据最新的人口普查,我们的人口增加了。

  8. Politically, the country is divided into fifty states.

    该国从行政上被划分为50个州。

9. The industrial area is centered largely in the north.

工业区大部分集中在北方。

10.The country is rich in natural resources. It has large quantity of mineral deposits.

该国自然资源丰富,有大量的矿藏。

11. This nation is noted for its economic stability.

该国以经济稳定而出名。

12. The U.S. is by far the biggest industrial country in the world.

到目前为止,美国是世界上最大的工业国。

13. My home is in the capital. It's a cosmopolitan city.

我的家在首都。它是一个国际大都会。

14. Geographically, this country is located in the southern hemisphere.

从地理位置上讲,这个国家位于南半球。

15. Britain is an island country surrounded by the sea.

英国是一个四周环海的岛国。

16. It's a beautiful country with many large lakes.

这是一个有着若干大湖的美丽国度。

17. Thames River is the second longest and the most important river in Britain.

泰晤士河是英国第二大河,也是英国最重要的河。

18. This part of the country is very mountainous.

这个国家的这部分土地被众多的山脉覆盖。

19. The land in this region is dry and parched.

这个地区的土地十分干旱。

20 Along the northern coast there are many high cliffs.

北海岸多危壁断崖。

21. There are forests here and lumbering is important.

此地多森林,故以伐木业为主。

22. In Brazil, many ancient forests are very well preserved.

在巴西,古老的森林被保存得十分完好。

23. The scenery is beautiful in the small islands in the Pacific Oceans.

太平洋上一些小岛的景色十分优美。

24. This mountain range has many high peaks and deep canyons.

此山高峰深谷众多。

25. What kind of climate do you have? Is it mild?

你们那里气候怎么样? 温和吗?

26. Britain has a maritime climate—winters are not too cold and summers are not too hot.

英国属于海洋性气候,冬季不过于寒冷,夏季不过于炎热。

27. How far is it from the shore of the Atlantic to the mountains?

从大西洋海岸到山区有多远?

28. Lumbering is very important in some underdeveloped countries.
在一些发展中国家,伐木业十分重要。

29. What's the longest river in the United States?
美国最长的河是什么河?

30. Are most of the lakes located in the north central region?
大部分湖泊是不是在北部的中心地区?

31. As you travel westward, does the land get higher?
你去西部旅行时,是不是地势越来越高?

32. The weather is warm and sunny here. Do you get much rain?
这里的天气温暖而晴朗。雨多吗?

33. Because of the warm and sunny weather, oranges grow very well here.
因为这里气候温暖,光照充足,橘子长势很好。

34. In this flat country, people grow wheat and corn and raise cattle.
这个国家地势平坦,人们种植小麦、玉米,饲养牲畜。

35. The ground around here is stony and not very good for farming.
这周围的土地多石,不适合耕种。

36. Is the coastal plain good for farming?
这种海边的平原有利于发展农业吗?

37. Is the plain along the river good for farming?
河畔的平原易于发展农业吗?

38. What are the principal farm products in this region?
这个地区的主要农产品是什么?

39. Milk, butter and cheese are shipped here from the dairy farms.
牛奶、黄油、奶酪都从奶制品农场运到这里。

40. They had to cut down a lot of trees to make room for farms.
他们不得不砍伐一些树木,从而为农场提供足够的空间。

41. At this time of the year farmers plow their fields.
一年中的这个时候农民们会耕种自己的土地。

42. On many farms you'll find cows and chickens.
在许多农场你都能发现奶牛和鸡。

43. If you have cows, you have to get up early to do the milking.
如果你有奶牛,你得早起挤牛奶。

44. Tractors have revolutionized farming.
拖拉机使农业发生了革命性的变化。

45. In the United States, there are many factories for making cloth.
美国有很多制布厂。

46. Factories employ both male and female workers.
工厂既雇用男工,也雇用女工。

47. If you work in a factory, you usually have to punch a clock.

如果你在工厂工作的话,你就得打卡。

48. Is meat packing a big industry in your country?

肉类加工在你们国家是不是一个大的产业?

49. Is it true that manufacturing of automobiles is a major industry?

汽车制造业是一个主要产业,是吗?

50. Mount Tai is situated in the western Shandong Provience.

泰山位于山东省西部。

Specialized Sentences

1. There are mandatory for use in works contracts which are estimated to cost more than $10 million.

对于成本估计超过 1,000 万美元的工程合同,必须使用该标准文件。

2. The corresponding substitution or modification of the bid must accompany the respective written notice.

投标文件的替换和修改必须附有相应的书面通知。

3. The bid security of the successful bidder shall be returned as promptly as possible once the successful Bidder has signed the Contract and furnished the required performance security.

一旦中标人签订了合同并按规定提交履约保证金后,中标人的投标保证金将尽快退还。

4. The employer shall open the bids in public in the presence of bidders' designated representatives at the address, date and time specified in the bidding document.

招标人应该按照招标文件规定的时间和地点在投标人指定代表出席的情况下公开开标。

5. Items against which no rate or price is entered by the Bidder will not be paid for by the Employer when executed.

投标人没有填入单价和价钱的项目在实施过程中招标人将不予支付。

6. It is undesirable that information relating to the examination, clarification and evaluation of bids before the award of a contract to the successful bidder is announced.

在中标者成功签约前就公布有关检查、清标、评标等的建议是不可取的。

7. The bidding documents should state clearly whether contracts will be awarded on the basis of unit prices or of a lump sum of the contract.

招标文件应明确说明合同是否将会以单价或合同总价为基础签订。

8. The size and scope of individual contracts will depend on the magnitude, nature, and location of the project.

单个合同的规模和范围将取决于项目的大小、性质和位置。

9. For projects requiring a variety of works and equipment such as power, water supply, or industrial projects, separate contracts are normally awarded for the civil works.

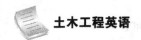

对于需要不同工程和设备(如电力、给排水或工业设施)的项目,通常会签订单独的土木工程合同。

10. For a project requiring similar but separate civil works or items of equipment, bids should be invited under alternative contract options.

对于需要类似但不同的土木工程或设备单元的项目,应该在备选合同方案下招标。

11. Contractors or manufacturers, small and large, should be allowed to bid for individual contracts or for a group of similar contracts at their option.

承包商或制造商,无论是小型还是大型的,应该被允许他们自主投标单个合同或一组类似的合同。

12. All bids and combinations of bids should be opened and evaluated simultaneously.

所有标书和组合投标都应该被同时公开并同步评估。

13. Detailed engineering of the works or goods to be provided should precede the invitation to bid for the contract.

所提供的工程或货物的详细的工程量都应该先于合同的邀请函发出。

14. In the case of turnkey contracts or contracts for large complex industrial projects, it may be undesirable to prepare technical specifications in advance.

对于交钥匙合同或大型的复杂的工业项目,事先准备技术规范可能不适当。

15. In such a case, it will be necessary to use a two-step procedure inviting un-priced technical bids subject to technical clarification and adjustments.

在这种情况下,它将需要使用一个两步骤程序进行不报价投标,并随时对其进行述标和调整。

16. Then, it is followed by the submission of priced proposals.

接下来提交有价格的投标。

17. The time allowed for preparation of bids should depend on the magnitude and complexity of the contract.

投标所需要的准备时间应取决于合同的规模和复杂程度。

18. Generally, not less than 45 days from the date of invitation to bid should be allowed for international bidding.

一般来说,对于国际招标,从招标邀请开始不少于 45 日。

19. Prospective bidders should be allowed to conduct investigations at the site before they submit their bids.

潜在投标人在投标前应能到施工现场进行调查。

20. The date, hour, and place for latest delivery of bids by the bidder and of the bid opening, should be announced in the invitation for bids.

投标人最迟提交标书的日期、时间、地点和开标的时间,应在招标须知中说明。

21. All bids should be opened at the stipulated time.

所有的投标应在规定的时间内开标。

22. Bids delivered after the time stipulated should be returned unopened.

在规定时间后提交的标书应退回,不能开标。

23. Unless the delay was not due to any fault of the bidder and its late acceptance would not give him any advantage over other bids.

除非过期的投标不是投标人的过错造成的才能被接受,但也要保证其不能获得超过其他投标的优势。

24. Bids should normally be opened in public.

通常应公开开标。

25. The name of the bidder and total amount of each bid, and any alternative bids should, when opened, be read aloud and recorded.

投标人的名字、投标的总金额以及其他备选标书被打开时,应当被大声读取和记录。

26. Extension of validity of bids should normally not be requested.

一般不应要求延长投标有效期。

27. If, in exceptional circumstances, an extension is required, it should be requested of all bidders before the expiration date.

如果在特殊情况下需要延期的,应在到期日期前告知所有投标人。

28. Bidders should have the right to refuse to grant all extension without forfeiting their bid bond.

投标人有权拒绝同意所有延长而不丧失其投标保证金。

29. Those who are willing to extend the validity of their bid should be neither required nor permitted to modify their bids.

不能要求也不能允许那些愿意延长其投标有效期的投标人修改其投标。

30. No bidder should be permitted to alter his bid after bid has been opened.

在开标后没有人可以修改投标。

31. Only clarifications not changing the substance of the bid may be accepted.

只有不改变标书实质的说明才能被接受。

32. The borrower may ask any bidder for a clarification of his bid but should not ask any bidder to change the substance or price of his bid.

借方可以要求投标人说明他的投标,但不应要求任何投标人修改标书或改变他们的出价。

33. Following the opening, it should be ascertained whether material errors in computation have been made in the bids.

开标后应查明是否在竞标中有重大计算错误。

34. If a bid is not substantially responsive to the bidding documents, or contains inadmissible reservations, it should be rejected.

如果投标在实质上不符合招标文件,或包含不能允许的保留,就要加以拒绝。

35. Unless it is alternative bid permitted, or requested, under the bidding documents.

除非它是根据投标文件所要求的或允许的替代投标。

36. A technical analysis should then be made to evaluate each responsive bid and to enable

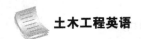

bids to be compared.

要进行技术分析,然后对每个投标进行评价、比较。

37. A detailed report on evaluation and comparison of bids should set forth the specific reasons on which the decision for the award of the contract.

应当编写一份详细的投标报告,评价和比较各个投标,据此决定授予合同。

38. The award of a contract should be made within the period specified for the validity of bids.

合同应当在规定的投标有效期内被授予。

39. The award of a contract should be made to the bidder whose responsive bid has been determined to be the lowest evaluated bid.

合同授予以最低价投标而且有能力的投标人。

40. The award of a contract should be made to the bidder who meets the appropriate standards of capability and financial resources.

合同授予财务方面符合相应标准的投标人。

41. Tender and bid activities shall conform to the principles of openness, fairness, impartialiey and good faith.

招投标活动应该遵循公开、公平、公正和诚实信用的原则。

42. The quantity of structural material required per square foot of floor of a high-rise building is in excess of that required for a low-rise building.

高层建筑单位面积楼层所需结构材料比低层建筑要多。

43. The vertical components carrying the gravity load, such as walls, columns, and shafts, will need to be strengthened over the full height of the building.

需要在整个高度范围内提高支撑重力荷载的竖向构件(如墙、柱和电梯井)的承载力。

44. Quantity of materials required for resisting lateral forces is even more significant.

但抵抗侧向力所需的材料用量甚至会更大。

45. The weight of structural steel in pounds per square foot of floor increases as the number of stories increases.

单位楼层所需结构材料用量随楼层数增加。

46. By using optimum structural systems with suitable width and arrangement, the additional material required for lateral force resistance can be controlled.

通过采用优化结构系统使其具有适当的广度和布局,可控制抵抗侧向力所需的额外材料。

47. Such that, even for buildings of 100 stories, the total structural weight of steel may be only about 34 pounds per square foot.

由此,甚至当建筑物层数高达100层时,单位面积总结构用钢量仅为每平方英尺约34磅。

48. Some buildings quite a bit shorter require much more structural steel.

一些比这低许多的建筑物却需要更多的结构钢件。

49. Remember that all high-rise buildings are essentially vertical cantilevers which are supported at the ground.

所有高层建筑实际上都是由地面上的竖向悬臂梁来支撑的。

50. Structurally desirable schemes can be obtained by walls, cores, rigid frames, tubular construction and other vertical subsystems to achieve horizontal strength and rigidity.

通过科学地设计具有水平强度和刚度的墙、核心筒、刚性框架、筒体结构和其他竖向体系，可以得到恰当的结构方案。

Task Two Sample Dialogue

Directions: *In this section, you are going to read several times the following sample dialogue about the relevant topic. Please pay special attention to five C's (culture, context, coherence, cohesion and critique) in the sample dialogue and get ready for a smooth communication in the coming task.*

(*The President of a local construction company discusses with Mr. Bian the possibility of forming a joint venture to bid for a project.*)

Daly: I was wondering whether it would be possible for us to form a joint venture to bid for...

Bian: For the Morpan Power Project?

Daly: You took the words out of my mouth, Mr. Bian! How did you know?

Bian: We're an international contractor, you know. We always keep our eyes open for new business opportunities.

Daly: This project is financed by the Inter-American Development Bank. If international bidders form joint ventures with domestic companies, they will be eligible for a 7.5 % margin of preference in the comparison of their bids with those of other bidders.

Bian: You've got a good point there. I like the idea, but can I have a detailed documentary introduction to your company? I have to report to our head office for approval.

Daly: Of course. I'll have it sent to you tomorrow. We should get to know each other better if we want to be working partners. I know that you're one of the top 225 contractors in the world with a good reputation. We're not one of those, but I think we're one of the best as well.

Bian: OK, Mr. Daly. As soon as this plan is approved by our head office, I'll let you know and we can have further discussions about our possible cooperation.

Daly: Thank you. I'm looking forward to your positive response.

(*After getting confirmation from the head office, Mr. Bian invites Mr. Daly again for a specific discussion on a joint venture agreement.*)

Bian: Now, Mr. Daly, we were talking last time about your idea on forming a joint venture for the Morpan Project. Our head office agrees to your proposal and I am authorized to

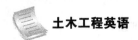

discuss it.

Daly: That's very good news, Mr. Bian. Do you have something specific in mind?

Bian: First, we have to make clear our purpose in this cooperation. That is, both sides agree to jointly prepare and submit a bid for the Morpan Project.

Daly: How much fund would you be prepared to provide for this purpose?

Bian: Are there any regulations regarding the foreign party's share in a joint venture in this country?

Daly: No, not really, but as a general practice, the leading company shall be the one which provides the most fund.

Bian: We'd like to act as the leader of the joint venture and lay out more than half of all the required fund for it, say, 55%.

Daly: That includes cash, construction equipment and other temporary facilities, if I understand correctly, Mr. Bian?

Bian: Of course.

Daly: I suppose you will provide the equipment and facilities used on your Project. How do we arrive at the true value of them?

Bian: That's a very good question. We could ascertain such values through joint assessment. In order to have an impartial assessment, we might also invite a third party to attend, if necessary.

Daly: That's good. So, we will provide 45% of the fund. I think I should be frank with you. We can provide the fund in local currency only, as we don't have foreign currencies.

Bian: Accordingly, you will share the profit, if any, in local currency only, too. The exchange rate for currency conversion will be the same as stated in the Owner's bidding documents.

Daly: I agree. All the profits, losses and liabilities arising out of the Contract shall be shared out between us in the same proportion as our respective fund contributions.

Bian: That's the basis of our agreement.

Task Three   Simulation and Reproduction

Directions: *The class will be divided into three major groups, each of which will be assigned a topic. In each group, some students may be the teacher, while others may be students. In the process of discussion, please observe the principles of cooperation, politeness and choice of words. One of the groups will be chosen to demonstrate the discussion to the class.*

1) Construction bidding in our daily life.

2) What factors should be paid attention to during the bidding stage?

3) The importance of bidding documents.

Task Four   Discussion and Debate

Directions: *The class will be divided into two groups. Please choose your stand in regard to the following controversy and support your opinions with scientific evidences. Please refer to the specialized terms and classical sentences in the previous parts of this unit.*

Joint ventures are a popular way to share the costs of bidding into new projects. It is temporary and is often dissolved or sold on completion of the project that brought the partners together. However, they will run a risk of failure because of compatibility and liability for partners' mistakes. How to solve these problems?

V. After-class Exercises

1. *Match the English words in Column A with the Chinese meaning in Column B.*

A	B
1) prequalification	A) 折旧
2) representation	B) 投标
3) depreciation	C) 开始
4) credit	D) 文件
5) document	E) 资格预审
6) bid	F) 授权
7) commencement	G) 竣工
8) completion	H) 附件
9) authorize	I) 代理人
10) attachment	J) 信誉

2. *Fill in the following blanks with the words or phrases in the word bank. Change the forms if it's necessary.*

bids	scope	magnitude	bond	substance
award	lump	linear	unopened	extension

1) It is undesirable that information relating to the evaluation of bids before the _____ of a contract to the successful bidder is announced.

2) The _____ of individual contracts will depend on the magnitude, nature and location of the project.

3) The bidding documents should state clearly whether contracts will be awarded on the basis of unit prices or of a _____ sum of the contract.

4) All _____ should be opened and evaluated simultaneously.

5) The _____ of the live load at any given time may be quite different from that specified by the building code.

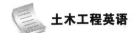

6）A structure is _____ if its response to loading is directly proportional to the level of the applied load.

7）_____ of validity of bids should normally not be requested.

8）Bidders should have the right to refuse to grant such all extension without forfeiting their bid _____.

9）Only clarifications not changing the _____ of the bid may be accepted.

10）Bids delivered after the time stipulated should be returned _____.

3. *Translate the following sentences into English.*

1）投标文件的替换和修改必须附有相应的书面通知。

_____

_____

2）在实施过程中,投标人没有填入单价和价钱的项目招标人将不予支付。

_____

_____

3）对于成本估计超过 1,000 万美元的工程合同,必须使用该标准文件。

_____

_____

4）招标人应该按照招标文件规定的时间和地点在投标人指定代表出席的情况下公开开标。

_____

_____

5）一旦中标人签订了合同并按规定提交履约保证金后,中标人的投标保证金将尽快退还。

_____

_____

4. *Please write an essay of about* 120 *words on the topic：**How to prepare bidding documents.** Some specific examples will be highly appreciated and watch out the spelling of some specialized terms you have learnt in this unit.*

_____

_____

_____

_____

_____

_____

_____

_____

_____

VI. Additional Reading

Main steps in the tender process

Each year, federal, state and local governments invite the private sector to submit competitive bids for the supply of goods and services. Government tender requests attract small, medium and large businesses across a wide range of sectors—from office supplies to major construction projects. Tender processes in the government sector share many common elements. What follows is an overview of the main steps you should take to prepare a competitive tender.

Attend tender information sessions. If you registered through a tender website, monitor the website for updates about the tender. Attend any tender information sessions offered. These are valuable opportunities to ask questions and make contact with the agency. They may also give you a chance to meet potential subcontractors or make contacts that could participate in a consortium(财团). Government agencies are also usually under no obligation to otherwise make available copies of information, presentations, etc., that may be given at an information session.

Review recent awarded contracts. If you are unclear about any requirements in the tender request, contact the tender coordinator to seek clarification. Also, review previously awarded contracts using the Queensland Contracts Directory. The details of recent government contracts over $10,000 that have been awarded are listed on tender websites.

Research your buyer. What type of companies won similar tenders in the past? What does the contracting agency look for? What can you do to match their expectations?

Write a compelling bid. Prepare your tender proposal. This includes planning, drafting and refining it. Make sure that you use the response forms provided and answer all questions. Stick to any word or page limits that there may be and (as a general rule) do not go altering things like fonts and font sizes and numbering unless expressly permitted.

Be clear about your structure and propositions. Decide on several key propositions you can use to set your tender apart from others. Review the evaluation criteria(标准) to gain a better understanding of what things the government agency is particularly looking for and will be evaluating your offer against. If you are not a strong writer, think about engaging a professional (a range of businesses offer tender writing services).

Understand the payment terms. When putting together your tender, make sure you are aware of the payment schedule specified. Government payment schedules will vary from agency to agency and procurement to procurement(采购).

You may not get paid as soon as the job is finished or goods are delivered. If you require payment different to that specified, you should detail this in your offer.

Present your bid. Tender panels responsible for high-value contracts may request a formal presentation from bidders. If you need to present your offer to an evaluation panel, stay focused

on the key messages in your proposal. Most importantly, prepare. Plan your presentation carefully, rehearse and, if you don't feel you're a strong presenter, get some coaching in presentation skills.

1. *Read the passage quickly by using the skills of skimming and scanning. And choose the best answer to the following questions.*

1) What should you do if you registered through a tender website? _____.

   A. Do nothing

   B. Cancel accounts

   C. Read more books

   D. Keep abreast of the latest information on tenders

2) What are the benefits of attending information meeting? _____.

   A. Hav a chance to ask question

   B. Communicate with institutions

   C. Join a consortium

   D. Meet potential subcontractors or make contacts

3) Government agencies have no obligation to do anything EXCEPT _____.

   A. provision of information by other means

   B. provision of information at information meetings

   C. copies of presentations at information meetings

   D. the above three options are correct

4) What should you do if you don't understand any of the requirements in the tender request? _____.

   A. Argu with the tender coordinator

   B. Contact the bid coordinator for clarification

   C. Refuse to bid

   D. Seek legal assistance

5) You may investigate your buyer about the following _____.

   A. What types of companies won similar bids in the past?

   B. What's the contract agency looking for?

   C. What can you do to meet their expectations?

   D. The above three options are correct.

6) What should be included in a convincing offer? _____.

   A. The company's influence

   B. The history of the company

   C. Preparation of your bid.

   D. Do nothing.

7) If you're not a powerful writer, what will you do when you're writing a bid? _____.

A. Write the bid by yourself

      B. Seek Professional Bid Writing Service

      C. Plagiarize of tenders

      D. Leave it to your subordinates

8) What should you do if the payment you request is different from that stipulated? _____.

      A. You should comply with the terms of the contract

      B. You should give up remuneration

      C. You should mark in the offer

      D. You should consult lawyers

9) What shall you do when you get the bid? _____.

      A. Start the process immediately

      B. Make consultation with partners

      C. Make sure you know the specified payment schedule

      D. Ignore the government's timetable

10) What should you do if you need to present your recommendations to the Panel? _____.

      A. You should make formal statements

      B. You should get some guidance on speaking techniques

      C. You should focus on the key messages in your proposal

      D. You should plan your speech carefully

2. *In this part, the students are required to make an oral presentation on either of the following topics.*

1) The features of bidding documents.

2) The types and function of tenders.

习题答案

Unit Three   Golden Gate Bridge

I. Pre-class Activity

Directions: *Please read the general introduction about **Golden Gate Bridge** and tell something more about the bridge to your classmates.*

Golden Gate Bridge

The Golden Gate Bridge is a suspension bridge spanning the Golden Gate, the one-mile-wide

(1.6 km) strait connecting San Francisco Bay and the Pacific Ocean. The structure links the American city of San Francisco, California—the northern tip of the San Francisco Peninsula—to Marin County, carrying both U.S. Route 101 and California State Route 1 across the strait. The bridge is one of the most internationally recognized symbols of San Francisco, California and the United States. It has been declared one of the Wonders of the Modern World by the American Society of Civil Engineers.

The Frommer's travel guide describes the Golden Gate Bridge as "possibly the most beautiful, certainly the most photographed, bridge in the world." At the time of its opening in 1937, it was both the longest and the tallest suspension bridge in the world, with a main span of 4,200 feet (1,280 m) and a total height of 746 feet (227 m).

II. Specialized Terms

Directions: *Please memorize the following specialized terms before the class so that you will be able to better cope with the coming tasks.*

arch bridge 拱形桥   beam bridge 梁式桥

38

single-log bridge 独木桥

viaduct bridge 高架桥

bowstring arch 系杆拱桥

box girder bridge 箱梁桥

cable-stayed bridge 斜拉桥

cantilever bridge 悬臂桥

clapper bridge 木板桥

covered bridge 廊桥(有顶盖)

acquaint vt. 使了解,使熟悉

abutment n. 桥墩, 桥基

acceleration n. 加速度

accessory adj. 附加的,辅助的

adjacent adj. 邻近的

aeroelastic adj. 空气弹性变形的

A-frame A 型骨架

aggregate v.使聚集

alternatively adv.要不,或者

analysis n. 分析,分解,解析

angle n. 角,角度,视角

angular adj. 有角的

application n. 请求,应用

arch n. 拱

architecture n. 建筑学

amplify v.扩大

battery acid level 电池酸位

battery cell volt 蓄电池电压

battery charger 电池充电器

battery electric locomotive 电力机车

battery electrolyte 电池电解液

battery n. 蓄电池

battery-powered device 电池推动装置

beacon n. 闪光指示灯

bead n. 珠子

bearing capacity 承载力

bearing force 承重能力

bearing pad 桥梁支座,轴承重

bearing pile 支承柱

bearing pin 轴承销

bearing plate 支承板

bearing stress 支承应力

bearing surface 支承面

bearing n. 支座

bedding n. 底层,层理

bed-plate n. 座板

bedrock n. 基层岩

behavior n. 性能,状况

belisha beacon 斑马线灯,黄波灯

bellow pot 气囊,气囊筒

belt conveyor 带式输送机

belt guard 皮带护罩

belt tension 皮带拉力

belt n. 带,皮带

bend n. 弯角,弯位,路弯

bending force 弯曲力

bending stress 弯曲应力

bentonite n. 膨润土,膨土岩

berm channel 斜水平台渠

berth n. 停泊处,泊位

bevel n. 斜角,斜面

bias n. 偏移

caisson n. 防水框,潜水箱

calculation n. 计算

canyon n. 峡谷

cap vt. 覆盖顶端

clapper bridge 木板桥

cofferdam n. 围堰

constituent n. 成份

covered bridge 廊桥

crane n. 起重机,吊车

debris n. 碎片,残骸

derrick n. 转臂起重机,(尤指船上的)吊

　杆式起重机

dynamic adj. 动态的

exert vt. 尽(力), 施加(压力等)

false work n. 脚手架

flotation n. 悬浮

guardrail n. 护栏	phase n. 阶段,时期
hazardous adj. 危险的	portion n. 部分
heterogeneity n. 多相性, 异质性	prefabricate vt. 预制
hoist v.吊起,提升,拉高	provision n. (法律文件)规定
hollow pier 空心板	stripping formwork 拆模
hydraulic adj. 液压驱动的	support bar 支撑杆
induce vt. 引起,导致	surface finish 表面平整度
ingenuity n. 巧妙,精巧	symmetry n. 对称
install vt. 安装, 安置	technical data 技术参数
mechanic n. 力学	temperature control 温度控制
organic n. 有机物质	temperature difference 温差

III. Watching and Listening

Task One　Golden Gate Bridge（I）

视频链接及文本

New Words

sail v.航行	reflect v.反映, 反射
navy n. 海军	entire adj. 全部的
bay n. 海湾	inlet n. 水湾,入口
dusk n. 黄昏	vermillion n. 朱砂,朱砂(红)色

Exercises

　　1. *Watch the video for the first time and answer the following questions.*

　　　1）How long is the Golden Gate Bridge? _____.

　　　　A. 7,000 feet　　　　　　　　　B. 8,000 feet

　　　　C. 9,000 feet　　　　　　　　　D. 6,000 feet

　　　2）What is the color of the bridge? _____.

　　　　A. Orange red　　　　　　　　　B. Orange yellow

　　　　C. Orange vermillion　　　　　　D. Orange purple

　　　3）What is the color the U.S. Navy originally wanted it to be painted? _____.

　　　　A. White and black　　　　　　　B. Yellow and red

　　　　C. Black and yellow　　　　　　　D. Black and red

　　　4）What do people think the Golden Gate Bridge is named after? _____.

　　　　A. The color of its paint　　　　　B. The color of its ground

　　　　C. The color of its sky　　　　　　D. The color of its sea

　　　5）When does the water turn from blue to gold? _____.

　　　　A. The moon sets at night　　　　B. The sun rises in the morning

C. The moon rises at dusk   D. The sun sets at dusk

2. *Watch the video again and decide whether the following statements are true or false.*

1) People don't think the Golden Gate Bridge is named after the color of its paint. (　)

2) On a clear day, when the sun sets at dusk, the water turns from blue to black. (　)

3) The Golden Gate Bridge is perhaps the most famous bridge in the world. (　)

4) In actuality, the bridge is named for the color of the water in the bay. (　)

5) It is nearly 9,000 feet long, and above water its towers are 640 feet tall. (　)

3. *Watch the video for the third time and fill in the following blanks.*

The Golden Gate Bridge is perhaps the most famous bridge in the world. It is famous for its _____, _____ and color. It is nearly 9,000 feet long and above water its _____ are 740 feet tall. Its _____ color is _____ known as "orange vermillion". Thousands of ships _____ under it every year and _____ it or not, the U.S. Navy _____ wanted it to be _____ black and _____ so it would be easy to see.

4. *Share your opinions with your partners on the following topic for discussion.*

Can you use a few lines to list what's your understanding about building a bridge? Please use an example to clarify your ideas.

Task Two Golden Gate Bridge (II)

New Words

fog n. 雾;迷惑

moisture n. 水分,湿度

rust n. 铁锈

expose vt. 揭露,使曝光

rumor n. 谣言

Richmond n. 里士满(地名)

Oakland n. 奥克兰(美国加利福尼亚州西部城市)

视频链接及文本

Exercises

1. *Watch the video for the first time and answer the following questions.*

1) When warm air blows over cold water, what does it tend to create? _____.

    A. Fog   B. Rain

    C. Smoke   D. Snow

2) Why is the bridge continually being painted? _____.

    A. In order to protect the bridge from melt

    B. In order to protect the bridge from wet

    C. In order to protect the bridge from frozen

    D. In order to protect the bridge from rust

3) What does the Bay Bridge connect? _____.

    A. The large cities of Oakland on the East Bay and San Francisco in the west

    B. The large cities of Oakland on the East Bay and San Francisco in the north

    C. The large cities of Oakland on the West Bay and San Francisco in the south

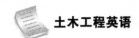

D. The large cities of Oakland on the West Bay and San Francisco on the east

4) Which bridge is beautiful to look at? _____.

   A. Richmond Bridge   B. The Bay Bridge

   C. None of them   D. The Golden Gate Bridge

5) Which bridge is a beautiful bridge to look from? _____.

   A. Richmond Bridge   B. The Golden Gate Bridge

   C. The Bay Bridge   D. None of them

2. *Watch the video again and decide whether the following statements are true or false.*

1) It is rumored that the bridge is being painted 365 days a year. (　)

2) San Francisco has large amounts of fog, especially in winter. (　)

3) The bay is always golden. (　)

4) When the warm California air meets the cold Atlantic Ocean, fog is formed. (　)

5) As soon as they finish painting it, they immediately start all over again. (　)

3. *Watch the video for the third time and fill in the following blanks.*

San Francisco has large _____ of fog, especially in _____. Fog is like a cloud close to the ground. When warm air _____ over cold water, it tends to _____ fog, so when the warm California air meets the cold _____ Ocean, fog is formed and the beautiful Golden Gate Bridge _____! Because of the fog, rain and _____ from the water, the bridge collects a lot of rust. Rust is the red _____ that collects on _____ after being _____ to moisture, or water. So in order to protect the bridge from rust, it is continually being painted. It is rumored that the bridge is being painted 365 days a year. As soon as they finish painting it, they immediately start all over again.

4. *Share your opinions with your partners on the following topic for discussion.*

    Do you know some famous bridges in China? Could you do some research and make a presentation about one of them?

IV. Talking

Task One　Classical Sentences

Directions: *In this section, some popular sentences are supplied for you to read and to memorize. Then, you are required to simulate and produce your own sentences with reference to the structure.*

General Sentences

1. —What was the weather like yesterday?

   —Yesterday it rained all day.

   —昨天天气怎样?

—昨天下了一天的雨。

2.—What will the weather be like tomorrow?

—It's going to snow tomorrow.

—明天天气怎样?

—明天有雪。

3. Do you think it's going to rain tomorrow?

你觉得明天会下雨吗?

4. I don't know whether it will rain or not.

我不知道明天会不会下雨。

5. It'll probably clear up this afternoon.

今天下午可能会放晴。

6. The days are getting hotter.

天气在变暖。

7. What's the temperature today?

今天多少度?

8. It's about twenty degrees centigrade this afternoon.

今天下午大约 20 摄氏度。

9. There's a cool breeze this evening.

今晚有股冷风。

10. Personally, I prefer winter weather.

就我个人而言,我比较喜欢冬天的天气。

11. What time are you going to get up tomorrow morning?

明天早上你打算几点起床?

12. I'll probably wake up early and get up at 6:30.

我可能会醒得比较早,大概 6 点半起床。

13. —What will you do then?

—After I get dressed, I'll have breakfast.

—接着你做些什么呢?

—穿上衣服后我就去吃早饭。

14. What will you have for breakfast tomorrow morning?

明天早餐你吃什么呢?

15. I'll probably have eggs and toast for breakfast.

早餐我可能会吃烤面包片和鸡蛋。

16. After breakfast, I'll get ready to go to work.

吃完早餐后我会准备上班。

17. I get out of bed about 7 o'clock every morning.

每天早上我 7:00 起床。

18. After getting up, I go into the bathroom and take a shower.

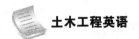

起床后我会去浴室冲个澡。

19. Then, I brush my teeth and comb my hair.
    然后,我刷牙梳头。

20. After brushing my teeth, I put on my clothes.
    我刷完牙后穿衣服。

21. After that, I go downstairs to the kitchen to have breakfast.
    然后,我下楼到厨房去吃早饭。

22. I'll leave the house at 8:00 and get to the office at 8:30.
    我 8 点离开家,8 点半到办公室。

23. I'll probably go out for lunch at about 12:30.
    我 12:30 左右去吃午饭。

24. I'll finish working at 5:30 and get home by 6 o'clock.
    我 5:30 下班,6:00 到家。

25. I'm always tired when I come home from work.
    下班以后我总是很累。

26. Are you going to have dinner at home tomorrow night?
    明天晚上你在家吃晚饭吗?

27. Do you think you'll go to the movies tomorrow night?
    明天晚上你会去看电影吗?

28. I'll probably stay home and watch television.
    我可能待在家看电视。

29. I'm not accustomed to going out after dark.
    我不习惯晚上出去。

30. When I get sleepy, I'll probably get ready to bed.
    当我感到困的时候,我就准备上床睡觉。

31. Do you think you'll be able to go to sleep right away?
    你觉得你现在就能去睡觉吗?

32. What do you plan to do tomorrow?
    明天你打算干什么?

33. I doubt that I'll do anything tomorrow.
    明天恐怕我什么也不做。

34. I imagine I'll do some work instead of going to the movies.
    我想我会工作,而不是去看电影。

35. Will it be convenient for you to explain your plans to him?
    你把你的计划跟他讲一下,方便吗?

36. What's your brother planning to do tomorrow?
    你哥哥明天计划去哪儿?

37. It's difficult to make a decision without knowing all the facts.

不知道全部事实而去做决定是很难的。

38. I'm hoping to spend a few days in the mountains.
    我想在山上待几天。

39. Would you consider going north this summer?
    你今年夏天想去北方吗?

40. If there's a chance you'll go, I'd like to go with you.
    如果你有机会去的话,我想和你一起去。

41. After you think it over, please let me know what you decide.
    在你仔细考虑之后,请告诉我你的决定。

42. Are you going to go anywhere this year?
    今年你打算去哪儿?

43. If I have enough money, I'm going to take a trip abroad.
    要是我有足够的钱,我打算出国旅行。

44. How are you going? Are you going by boat?
    你打算怎样去? 是不是乘船去?

45. It's faster to go by plane than by boat.
    坐飞机比坐船快。

46. What's the quickest way to get there?
    到那儿去最快的交通方式是什么?

47. Altogether it will take ten days to make the trip.
    这次旅行总共要花十天时间。

48. It was a six-hour flight(journey/travel/voyage).
    这是一次 6 小时的旅程。

49. I'm leaving tomorrow, but I haven't packed my suitcases yet.
    我明天出发,可是我的箱子到现在还没有整理好。

50. I hope you have a good time on your trip.
    祝你旅途愉快。

Specialized Sentences

1. The third party selected shall be acceptable to both sides.
   第三方的选择要经双方认可。

2. Please give us a copy of this information (technical specification, instruction, manual, document, diagram, catalog).
   请给我们一份这个资料(技术规程、说明书、手册、文件、图表、目录)的复印本。

3. Please send us a technical liaison letter about it.
   请给我们一份有关此事的技术联络单。

4. A working drawing must be clear and complete.
   工作图必须简明完整。

5. Data on equipment can be found in the related information.

设备的数据可从有关的资料中找到。

6. Have you any idea how to use the manufacturer's handbook.

你知道怎样使用这本厂家手册吗？

7. Public utilities are carefully regulated here.

公用事业在这里被严格的管理。

8. The pressure maintained in the water main is two kilogram per square centimeter.

自来水总管里的水压保持为 2 千克/平方厘米。

9. The water has been treated (softened), but it is not drinkable water.

这水经过处理(软化),但不是饮用水。

10. The common service voltage of electric power in our country is 220/380 volt.

我国普通供电电压为 220/380 伏。

11. There is a switch board mounted on the wall.

在墙上装有一个开关板。

12. We have an emergency-standby electric generator with a capacity of 300 kilowatts.

我们有一台 300 千瓦的应急备用发电机。

13. The substation equipped with a transformer of 500 kVA is at the south of the plant.

在工厂的南边有一座装有 500 千伏安变电器的变电站。

14. The pressure of the compressed air at the work site is about $7kg/cm^2$.

工地用压缩空气的气压约为 7 千克/平方厘米。

15. There is a steam heating system (air conditioning system) in the work-shop.

这车间里装有一个蒸汽加热系统(空调系统)。

16. Please give a description about this project.

请对这个工程项目作一个叙述说明。

17. Would you tell us the technical characteristic about this project?

你能告诉我们有关这个工程项目的技术特性吗？

18. The works can't be deemed to be substantially completed until you have presented them to us.

贵方将工程移交给我们之前,不能被认为基本上竣工了。

19. Everyone is very pleased to learn that our company has won the contract.

得知我们公司中标,大家都很高兴。

20. This contract is made in two orignials that should be beld by each party.

此合同一式二份,由双方各执一正本。

21. You may contact the receptionist if you want to make a long distance call (to make overseas call, to send an overseas telegram).

如果你要打长途电话(海外电话、国际电报),可以与前台联系。

22. The cargo vessel docked at wharf number 5 yesterday afternoon.

货船昨天下午停泊在 5 号码头。

23. There is a freeway (main highway) from here to Nanjing.

从这里到南京有一条高速公路(主要公路)。

24. Welcome to our construction site.

欢迎来到我们工地。

25. It is very simple and crude here. Do not mind, please.

这里很简陋,请别介意。

26. I wonder whether the emergency generator could start automatically in case the main generator breaks down.

我想知道在主发电机发生故障时,应急发电机是否能自动启动。

27. I am a site engineer (director, workshop head, chief of section, foreman, worker, staff member).

我是一名现场工程师(厂长、车间主任、班组长、领工、工人、职员)。

28. Mr. Wang is responsible for this task.

王先生负责这项工作任务。

29. Here is our engineering office (drawing office, control room, laboratory, meeting room, common room and rest room).

这是我们的工程技术办公室(绘图室、调度室、实验室、会议室、休息室、洗手间)。

30. I wonder if you could affect the shipment early next month.

我想知道你们是否能在下个月上旬装船。

31. I'm very pleased to learn that the works have passed all the completion tests.

我很高兴地得知工程通过了全部竣工试验。

32. The shift will start at half past seven a.m.

早班从 7 点半开始。

33. All has gone well with our site work plan.

一切均按照我们的现场工作计划进行。

34. We have flexible work hours during the summer.

我们在夏季施行弹性工作制。

35. It would only incur greater losses if we went on like this.

这样干下去只能导致更大损失。

36. Put on your safety helmet, please.

请戴上安全帽!

37. Danger! Look out! Get out of the way.

危险! 当心! 快躲开!

38. Here is our pipe prefabrication workshop (steel structure fabrication shop, machine shop, boiler room, air compressor station, concrete mixing unit).

这里是我们的管道预制车间(钢结构制作厂、机械加工车间、锅炉房、空气压缩机站、混凝土搅拌站)。

39. Would you like to see this process (machine)?

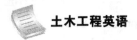

你要看看这工艺(机器)吗？

40. Would you like to talk to the welder (inspector)?

你要和焊工(检查员)谈谈吗？

41. The factory (work shop, equipment) produces pipe fittings (spare parts, fasteners).

这工厂(车间、设备)生产管件(配件、紧固件)。

42. I am sorry, and do not touch this, please.

很抱歉,请勿触碰这里!

43. Smoking and lighting fires are strictly forbidden at here.

这里严禁烟火。

44. Look at the sign："Danger! Keep out!"

注意标牌:危险勿进!

45. There is a temporary facility for site brickwork (wood work, ironwork, paintwork).

这是一个现场制砖(木制品、铁制品、油漆制品)的临时设施。

46. Let me show you around and meet our workers.

让我带你走一圈,并见见我们的工人。

47. We would like to know your opinion about our site work.

我们想听取你对我们现场工作的意见。

48. This item of work is deemed to have been included in the Contract Price and you're not entitled to an additional payment.

这项工作已包括在合同价格中,你们无权获得额外的付款。

49. Some training will fit them for the job.

经过一些训练,他们就能胜任这项工作。

50. By the end of this month, we shall have carried out our plan.

到这个月月底,我们将实现我们的计划。

Task Two   Sample Dialogue

Directions：*In this section, you are going to read several times the following sample dialogue about the relevant topic. Please pay special attention to five C's (culture, context, coherence, cohesion and critique) in the sample dialogue and get ready for a smooth communication in the coming task.*

(*Mr. Yang from COCC has made an appointment by telephone with Mr. Howell, President of World Shelters, a local company which specializes in residential buildings. They are going to talk about the possibility of subletting the construction camp to World Shelters.*)

Receptionist：Good morning, sir. Can I help you?

Yang：   Good morning. I have an appointment with Mr. Howell at 9 a.m.

Receptionist：Your name, sir?

Yang: Yang, from COCC, the Hydro.

Receptionist: Ah, yes. Mr. Howell is expecting you in his office. Let me show you in. This way, please.

Yang: Thank you.

(*Entering Mr. Howell's office*)

Howell: Hello, Mr. Yang. Very pleased to meet you in person.

Yang: Very pleased to meet you, too, Mr. Howell.

Howell: Take a seat, please.

Yang: Thank you.

Howell: Well, how is everything going, Mr. Yang?

Yang: Not too bad. Just busy. We're quite new here, you know.

Howell: What's your first impression of our country?

Yang: It's a beautiful country with plenty of primeval forests. People here are friendly and very easy to get along with. It seems for me that most of the people here speak quite different English. It's hard to understand them sometimes.

Howell: You will soon get used to it. Only officials, teachers and businessmen speak standard English.

Yang: So it seems.

Howell: Let's get down to business, Mr. Yang. You asked me on the phone whether we would like to bid for a construction camp to accommodate eighty men. Could you put it in more detail, please?

Yang: As you know, the camp will mainly be for the accommodation of about eighty Chinese men who will be working for the Hydro. It includes double occupancy staff units, four man occupancy units for workers and cooks, a kitchen and dining unit, bath units, office units, a conference room and a recreation unit.

Howell: So the camp will accommodate about eighty men. Do you have any specific requirements?

Yang: I would like to hear your recommendation.

Howell: In my opinion, there's a kind of prefabricated house which is most suitable for such a camp. It's easy and economical to transport, fast to erect and very convenient to dismantle for either relocation or disposal when the whole project is completed.

Yang: Sounds fine. What's it made of?

Howell: Light concrete slabs. They are often imported from a local supplier in El Salvador.

Yang: Good. Would you please give us a quotation for such a camp on a turn-key basis as soon as possible?

Howell: All right. Where will the camp be located?

Yang: Somewhere on Arenal Road in Benque Town, Cayo District. I'll let you know right

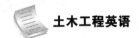

after we finalize it with the Owner.

Howell： Who will be responsible for leveling the ground for the camp site?

Yang： We will do that. You will be responsible for the water and power supply. One more thing, this is a duty-free project. So everything imported for it is duty-free. Please take it into consideration in your quotation.

Howell： In that case, our quotation will be much lower.

Yang： We really appreciate that.

Howell： When do you want the camp to be completed?

Yang： Within sixty days, starting from our notice to commence the work.

Howell： OK, Mr. Yang. I hope we can be given the opportunity to work for you and if so, we will hold ourselves responsible to you and I believe we will surely do a good job.

Yang： I hope so, too.

Task Three　Simulation and Reproduction

Directions： *The class will be divided into two major groups, each of which will be assigned a topic. In each group, some students may be the teacher, while others may be students. In the process of discussion, please observe the principles of cooperation, politeness and choice of words. One of the groups will be chosen to demonstrate the discussion to the class.*

1) How many kinds of bridges you learn from this unit? Which one impressed you most?

2) An important story related to bridges in your life.

Task Four　Discussion and Debate

Directions： *The class will be divided into two groups. Please choose your stand in regard to the following controversy and support your opinions with scientific evidences. Please refer to the specialized terms and classical sentences in the previous parts of this unit.*

On the Changjiang River, there are already several bridges. But the government still wants to build more bridges. Do you agree with the government or not? Why? Please think about it and give your reasons.

V. After-class Exercises

1. *Match the English words in Column A with the Chinese meaning in Column B.*

A	B
1) arch bridge	A) 高架桥
2) beam bridge	B) 独木桥
3) single-log bridge	C) 箱梁桥

4) viaduct bridge   D）系杆拱桥

5) bowstring arch   E）梁式桥

6) box girder bridge   F）斜拉桥

7) cable-stayed bridge   G）悬臂桥

8) cantilever bridge   H）木板桥

9) clapper bridge   I）廊桥(有顶盖)

10) covered bridge   J）拱形桥

2. *Fill in the following blanks with the words or phrases in the word bank. Change the forms if it's necessary.*

accept	date	capacity	vessel	forbid
regulate	board	pressure	flexible	opinion

1) The third party selected shall be _____ to both sides.

2) Public utilities are carefully _____ here.

3) We have _____ work hours during the summer.

4) We have an emergency-standby electric generator with a _____ of 300 kilowatts.

5) We would like to know your _____ about our site work.

6) Smoking and lighting fires are strictly _____ at here.

7) The _____ of the compressed air at the work site is about $7kg/cm^2$.

8) _____ on equipment can be found in the related information.

9) The cargo _____ docked at wharf number 5 yesterday afternoon.

10) There is a switch _____ mounted on the wall.

3. *Translate the following sentences into English.*

1) 请给我们一份有关此事的技术联络单。

2) 我想知道在主发电机发生故障时,应急发电机是否能自动启动。

3) 贵方将工程呈现给我们之前,不能被认为基本上竣工了。

4) 我很高兴地得知工程通过了全部竣工试验。

5) 设计图必须简明完整。

4. *Please write an essay of about 120 words on the topic: Importance of Bridges in our Life. Some specific examples will be highly appreciated and watch out the spelling of some specialized terms you have learnt in this unit.*

_____

_____

_____

_____

_____

_____

VI. Additional Reading

Golden Gate Bridge

[A] Before the bridge was built, the only practical short route between San Francisco and what is now Marin County was by boat across a section of San Francisco Bay. Ferry(摆渡) service began as early as 1820, with regularly scheduled service beginning in the 1840s to transport water to San Francisco.

[B] The Sausalito Land and Ferry Company service, launched in 1867, eventually became the Golden Gate Ferry Company, a Southern Pacific Railroad subsidiary(子公司), the largest ferry operation in the world by the late 1920s. Once for railroad passengers and customers only, Southern Pacific's automobile ferries became very profitable and important to the regional economy. The ferry crossing between the Hyde Street Pier in San Francisco and Sausalito in Marin County took approximately 20 minutes and cost $1.00 per vehicle, a price later reduced to compete with the new bridge. The trip from the San Francisco Ferry Building took 27 minutes.

[C] Many wanted to build a bridge to connect San Francisco to Marin County. San Francisco was the largest American city still served primarily by ferry boats. Because it did not have a permanent link with communities around the bay, the city's growth rate was below the national average. Many experts said that a bridge could not be built across the 6,700 ft. (2,042 m) strait, which had strong, swirling tides and currents, with water 500 ft. (150 m) deep at the center of the channel and frequent strong winds. Experts said that ferocious(凶猛的) winds and blinding fogs would prevent construction and operation.

[D] Although the idea of a bridge spanning the Golden Gate was not new, the proposal that eventually took hold was made in a 1916 San Francisco Bulletin article by former engineering student James Wilkins. San Francisco's City Engineer estimated the cost at $100 million, which would have been $2.12 billion in 2009 and impractical for the time. He asked bridge engineers whether it could be built for less. One who responded, Joseph Strauss, was an ambitious engineer and poet who had, for his graduate thesis, designed a 55-mile-long (89 km) railroad bridge across the Bering Strait.

[E] At the time, Strauss had completed some 400 drawbridges—most of which were inland—and nothing on the scale of the new project. Strauss's initial drawings were for a massive cantilever(悬臂梁) on each side of the strait, connected by a central suspension segment, which Strauss promised could be built for $17 million.

[F] Local authorities agreed to proceed only on the assurance that Strauss would alter the design and accept input from several consulting project experts. A suspension-bridge design was considered the most practical, because of recent advances in metallurgy(冶金学).

[G] Strauss spent more than a decade drumming up support in Northern California. The bridge faced opposition, including litigation(诉讼), from many sources. The Department of War was concerned that the bridge would interfere with ship traffic. The navy feared that a ship collision or sabotage(蓄意破坏) to the bridge could block the entrance to one of its main harbours. Unions demanded guarantees that local workers would be favoured for construction jobs. Southern Pacific Railroad, one of the most powerful business interests in California, opposed the bridge as competition to its ferry fleet and filed a lawsuit against the project, leading to a mass boycott of the ferry service.

[H] In May 1924, Colonel Herbert Dewayne held the second hearing on the Bridge on behalf of the Secretary of War in a request to use federal land for construction. Dewayne, on behalf of the Secretary of War, approved the transfer of land needed for the bridge structure and leading roads to the "Bridging the Golden Gate Association" and both San Francisco County and Marin County, pending further bridge plans by Strauss. Another ally was the fledgling automobile industry, which supported the development of roads and bridges to increase demand for automobiles.

[I] The bridge's name was first used when the project was initially discussed in 1917 by M. M. O'Shaughnessy, city engineer of San Francisco, and Strauss. The name became official with the passage of the *Golden Gate Bridge and Highway District Act* by the state legislature in 1923, creating a special district to design, build and finance the bridge. San Francisco and most of the counties along the North Coast of California joined the Golden Gate Bridge District, with the exception being Humboldt County, whose residents opposed the bridge's construction and the traffic it would generate.

[J] Strauss was the chief engineer in charge of overall design and construction of the bridge project. However, because he had little understanding or experience with cable-suspension designs, responsibility for much of the engineering and architecture fell on other experts. Strauss's initial design proposal (two double cantilever spans linked by a central suspension segment) was unacceptable from a visual standpoint. The final graceful suspension design was

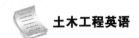

conceived and championed by Leon Moisseiff, the engineer of the Manhattan Bridge in New York City.

[K] Irving Morrow, a relatively unknown residential architect, designed the overall shape of the bridge towers, the lighting scheme, and Art Deco elements, such as the tower decorations, streetlights, railing, and walkways. The famous International Orange color was originally used as a sealant(密封胶) for the bridge. The US Navy had wanted it to be painted with black and yellow stripes to ensure visibility by passing ships.

1. *Read the passage quickly by using the skills of skimming and scanning. And choose the best letter standing for each paragraph above in response to the following sentences.*

1) San Francisco was the largest American city still served primarily by ferry boats.

2) The Department of War was concerned that the bridge would interfere with ship traffic.

3) A suspension-bridge design was considered the most practical, because of recent advances in metallurgy.

4) Strauss's initial drawings were for a massive cantilever on each side of the strait, connected by a central suspension segment, which Strauss promised could be built for $17 million.

5) The trip from the San Francisco Ferry Building took 27 minutes.

6) A ferry service began as early as 1820, with a regularly scheduled service beginning in the 1840s for the purpose of transporting water to San Francisco.

7) San Francisco's City Engineer estimated the cost at $100 million, which would have been $2.12 billion in 2009 and impractical for the time.

8) The famous International Orange color was originally used as a sealant for the bridge.

9) Strauss was chief engineer in charge of overall design and construction of the bridge project.

10) In May 1924, Colonel Herbert Deakyne held the second hearing on the Bridge on behalf of the Secretary of War in a request to use federal land for construction.

2. *In this part, the students are required to make an oral presentation on either of the following topics.*

1) The secrets of the Golden Gate Bridge.

2) The extraordinary design of the Golden Gate Bridge.

习题答案

Unit Four Construction Contract

I. Pre-class Activity

Directions: *Please read the general introduction about **Peter Schuyler Bruff** and tell something more about the great civil engineer to your classmates.*

Peter Schuyler Bruff

Peter Schuyler Bruff (1812~1900) was an English civil engineer best known for founding the seaside resort town of Clacton on Sea, Essex, and for improving the lives of residents in the Essex towns of Walton-on-the-Naze, Colchester and Harwich. Bruff was born in Portsmouth. While working with Eastern Counties Railway from Shoreditch to Colchester, he began work on the Chappel Viaduct, which was constructed between 1847 and 1849. The viaduct carries the Sudbury Branch Line across the Colne Valley in Essex. It stands 80 feet (24 m) above the river, has 32 arches and is 1,066 feet (325 m) long. The viaduct contains 4.5 million bricks. It was Bruff's dream for the line to Colchester to carry on as far as Ipswich but the railway company did not have sufficient funds. As a result, Bruff formed his own company, the Eastern Union Railway, and built the line himself, including the 361 yd. (330 m) tunnel through Stoke Hill by Ipswich railway station.

While working on the Ipswich line in 1855 he bought a house, Burnt House Farm, in Walton, an already established but unremarkable town on the Essex coast near Frinton. He began to work on developing Walton as a recognized seaside resort. He took a major step in accomplishing this when in 1867, having accomplished the Ipswich line, he built another railway line to Walton. Peter Bruff's pier at Walton replaced an existing smaller pier which was blown down by a storm in 1881. Bruff is regarded by some as the Richard Branson of the 19th Century for the work he did in Clacton, which was virtually non-existent when he arrived.

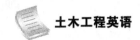

II. Specialized Terms

Directions: *Please memorize the following specialized terms before the class so that you will be able to better cope with the coming tasks*

accumulate v. 积累

accurate adj. 准确的,精确的

activate v. 启动

advance v. 前进,行进

advocate v. 拥护,提倡

affirm v. 断言,申明

aggregate n. (混凝土用的)骨料

allied adj. 结盟的

alter v. 改变,更改

alternate v. 交替,轮流

alternative n. 代替物

anchor n. 锚

anonymous adj. 匿名的

anticipate v.预期,预料

apart adv.分离

apparent adj. 显然的

appearance inspection 外观检查

appraise v.估量,估计

appropriate adj. 合适的,恰当的

approximate adj. 大概的

array v.排列

artificial adj. 人工的

ascend v.上升,升高

ascertain v.查明,弄清

ashore adv.向岸上,向陆地

assemble v.集合,召集

associate v.(使)联系

beforehand adj. 预先的,事先的

bending moment 弯曲度,弯矩

binding strength 黏结强度

block n. 立方体

boom n. 帆桁

bore v.钻孔,挖(通道)

brace n. 托架,支架

brittle adj. 易碎的

bump v.碰,撞

burst v.爆炸,爆裂

calibration n.标定

casual adj. 无计划的

cease v.停止,终止

certify v.证明,证实

character n. 性质,特质

compulsory adj. 义务的

conceive v.构想出,设想

concrete release sheet 混凝土预制板

construction joint 施工缝

contamination n. 污染

coverage n. 覆盖范围

crush v.压碎,彻底毁掉

curing agent 固化剂

delivery sheet 运料单

direct discharge 直接入仓

elevate v.举起,提高

elevation n. 高程,正视图

emigrate v.移居外国

emit v.发出,散发(光、热、电度等)

existing adj. 现有的,现存的

expansion joint 伸缩缝

facilitate v.使容易,使便利

final inspection 终验

fine aggregate 细骨料

gauge n. 测量仪器(或仪表)

geometry n. 几何结构

gimlet auger 钻,无柄钻

girder n. 大梁

gouge n. 圆凿，沟

graph n. 图表，图解，曲线图

hammer n. 锤子

hand drill 手钻

handsaw n. 手锯

homogeneous 各向同性

horizontal adj. 水平的

identification n. 鉴定

impact n. 碰撞，冲击力

layer（lift）height 层高

moment of inertia 惯性矩，转动惯量

apply v.应用，施加

rack v.架子，支架，搁架

regulate v.管理，控制

reserve v.保留，储备

resolve v.决定

resort v.求助，凭借，诉诸

reverse adj.相反的，颠倒的

revise v.修订，校订

revive v.苏醒，复苏

ridiculous adj. 荒唐的

rigid adj.刚性的，坚硬的

rot v.腐烂，腐坏

roundabout adj. 绕道的

scrape v.刮，擦

sector n. 扇形

secure adj. 安全的，可靠的

segregate v.分离，隔离

setback v.倒退

shabby adj. 简陋的，破旧的

shatter n. 碎片，粉碎

weight gravity 比重

spectrum n. 范围，幅度

theoretical elongation 理论伸长量

water-cement ratio 水灰比

III. Watching and Listening

Task One　The life of Building Contract Manager

视频链接及文本

New Words

residential adj. 住宅的

portion n. 一部分

structural adj. 结构的

concrete n. 混凝土

amenity n. 便利设施

multiple adj. 多重的

multiple n. 专家评价

schedule n. 进度表

prioritize v.划分优先顺序

maintain vt. 保持

budget n. 预算

Exercises

1. *Watch the video for the first time and choose the best answers to the following questions.*

　　1) We can conclude from the passage that the author was _____.

　　　　A. an engineer

　　　　B. a tour guide

　　　　C. a city constructor

　　　　D. a construction administrator

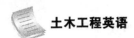

2) The office portion _____.

  A. has a similar structural system with the residential one

  B. has a totally different structural system than the residential one

  C. has no relationship with the residential one

  D. has the same structural system with the residential one

3) When will the project be finished? _____.

  A. After two years

  B. After two month

  C. Over two years

  D. About two month

4) In order to make our clients to be happy, we have done the following EXCEPT _____.

  A. listening to what our clients say

  B. replacing finishes based on some requests that we've done

  C. reducing common corridor finishes tourists goes

  D. making sure that they're aware of what we're substituting

5) The most challenging thing for the construction administration is _____.

  A. bringing a lot of expertise

  B. keeping everything coordinated

  C. maintaining the quality of the project

  D. solving all these problems within the time constraints

2. *Watch the video again and decide whether the following statements are true or false.*

 1) The construction administration team here has been working in London. (　　)

 2) The office is a true mixed-use project, 350,000 square feet.(　　)

 3) The project team is made up of multiple designers, multiple consultants.(　　)

 4) The most challenging thing for them was raising all these problems.(　　)

 5) They pay great attention to the quality, budget and the schedule.(　　)

3. *Watch the video for the third time and fill in the following blanks.*

This is really a true mixed-use project, 650,000 square feet of office and 350,000 square feet of _____. It's a true vertical _____. We essentially have two _____ stacked on top of each other. There is the office _____ which actually has a completely different _____ system than the residential does, so we go from _____ building to a _____ building separated by a _____ floor and the amenity floors. So we've been working on construction _____ for over two years now on that _____ and we're probably about two months away right now. This is a team that gets to bring it home. We're all excited. We see the light at the end of the now.

4. *Share your opinions with your partners on the following topic for discussion.*

 1) How do you feel the day in the life of a contract manager? Please summarize the features of project management.

 2) Can you use a few lines to list what's your understanding about contract management?

Please use an example to clarify your thoughts.

Task Two   Interview with Clark

New Words

contract n. 合同

services n. 公共事业机构(公司)

operative n. 工作人员

divination n. 预言

qualification n. 资格

stonemason n. 石工

trainee n. 实习生

foreman n. 工头

视频链接及文本

Exercises

1. *Watch the video for the first time and choose the best answers to the following questions.*

1) What does an operative should do? _____.

   A. Make sure the jobs are done on time   B. Divinate life

   C. Try and gets yourself bawled   D. Read somewhere

2) The word divination means? _____.

   A. Divine   B. Respectful

   C. Prophecy   D. Meaningful

3) What was Marie Clark's attitude toward sports? _____.

   A. It was formidable   B. It was very good

   C. It was healthy   D. It was joyful

4) How does the speaker become a formidable Jack? _____.

   A. Make sure things that are getting done properly

   B. Rate through the rank

   C. Play sports

   D. Fall back on the other things you'll know certainly looking towards the future.

5) Which style does this video belong to? _____.

   A. Prose   B. Short story

   C. Argumenttion   D. Declarative

2. *Watch the video again and decide whether the following statements are true or false.*

1) Marie Clark ended up as a contract manager.(   )

2) He tried the construction industry as a mason. (   )

3) He started this job as a project management in 1983.(   )

4) He basically started with an internship Jack.(   )

5) He disliked taking exercise in spite of high salary.(   )

3. *Watch the video for the third time and fill in the following blanks of the table.*

I tried the _____ trades as our stonemason. It was a good _____ job and I knew people that were stonemasons and I sort of knew that the KPA _____ he hired, I thought it's going to be a good job. In 1985, I, sort of what basically from a trainee steeple Jack, rate through the

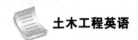

_____, became a formidable jack. And I do enjoy playing _____ and all kinds of sports I saw that I was pretty good. I really enjoy doing and still enjoyed doing sports. Obviously I know about the _____ so you've just got to fall back on the other things. You'll be _____ looking towards the _____. To become an a _____ foreman who just sort of travels, making sure things that are getting done _____.

4. *Share your opinions with your partners on the following topic for discussion.*

　　1）Do you know how to prepare a contract?

　　2）Can you design a contract with the aid of computer?

IV. Talking

Task One　Classical Sentences

Directions：*In this section, some popular sentences are supplied for you to read and to memorize. Then, you are required to simulate and produce your own sentences with reference to the structure.*

General Sentences

1. Do you really want to know what I think?
   你真想知道我在想什么吗？

2. Please give me your frank opinion.
   请告诉我你真实的想法。

3. Of course I want to know what your opinion is.
   当然，我很想知道你的看法。

4. What do you think? Is that right?
   你认为怎样？是对的吗？

5. Certainly. You're absolutely right about that.
   当然，你的看法完全正确。

6. I think you're mistaken about that.
   我想你弄错了。

7. I'm anxious to know what your decision is.
   我很想知道你的决定。

8. That's a good/great/fantastic/excellent idea.
   这个想法很好。

9. In my opinion, that's an excellent idea.
   我认为这是个好主意。

10. I'm confident you've made the right choice.
    我相信你做了个正确的决定。

11. I want to persuade you to change your mind.
    我想劝你改变主意。

12. Will you accept my advice?
    你会接受我的建议吗?

13. He didn't want to say anything to influence my decision.
    他不想说任何话来影响我的决定。

14. She refuses to make up her mind.
    她不肯下决心。

15. I assume you've decided against buying a new car.
    我想你已经决定不买新车了吧。

16. It took him a long time to make up his mind.
    他用了很久才下定决心。

17. You have your point of view and I have mine.
    你有你的观点,我也有我的想法。

18. You approach it in a different way from mine.
    你处理这件事的方式和我不一样。

19. I won't argue with you, but I think you're being unfair.
    我不想和你争辩,但是我认为你是不公平的。

20. That's a liberal point of view.
    那是一种自由主义的观点。

21. He seems to have a lot of strange ideas.
    他好像有很多奇怪的想法。

22. I don't see any point in discussing the question any further.
    我没有任何必要再进一步讨论这个问题。

23. What alternatives do I have?
    我还有什么方法?

24. Everyone is entitled to his own opinion.
    每个人都有自己的观点。

25. She doesn't like anything I do or say.
    无论我做什么说什么,她都不喜欢。

26. There are always two sides to everything.
    任何事物都有两面性。

27. We have opposite views on this.
    关于这个问题,我们有不同的观点。

28. Please forgive me. I didn't mean to start an argument.
    请原谅。我并不想引起争论。

29. I must know your opinion. Do you agree with me?
    我必须了解你的想法。你同意吗?

30. What points are you trying to make?

   你想表达什么观点?

31. Our views are not so far apart, after all.

   毕竟我们的意见没有多大分歧。

32. We should be able to resolve our differences.

   我们应该能解决我们的分歧。

33. If you want my advice, I don't think you should go.

   如果你想征求我的意见,我认为你不应该去。

34. I suggest that you tear up the letter and start over again.

   我建议你把信撕掉,再重新写一遍。

35. It's only a piece of suggestion and you can do what you want.

   这只是一个建议,你可以做你想做的。

36. Let me give you a little advice.

   让我给你提点建议。

37. If you don't like it, I wish you would say it.

   如果你不喜欢,我希望你能说出来。

38. Please don't take offense. I only wanted to tell you what I think.

   请别生气。我只是想告诉你我的想法。

39. My feeling is that you ought to stay home tonight.

   我觉得你今晚应该待在家里。

40. It's none of my business, but I think you ought to work harder.

   这不关我的事,但我认为你应该更努力地工作。

41. In general, my reaction is favorable.

   总体来说,留给我的印象是好的。

42. If you don't take my advice, you'll be sorry.

   如果你不听我的劝告,你会后悔的。

43. I've always tried not to interfere in your affairs.

   我总是尽量不干涉你的事情。

44. I'm old enough to make up my own mind.

   我已经大了,可以自己做决定了。

45. Thanks for the advice, but this is something I have to figure out myself.

   谢谢你的建议,但这是我必须自己解决的问题。

46. He won't pay attention to anybody. You're just wasting your breath.

   他谁的建议都不会听。你们是在白费口舌。

47. You can go whenever you wish.

   你愿意什么时候去就什么时候去。

48. We're willing to accept your plan.

   我们愿意接受你的计划。

49. He knows it's inconvenient, but he wants to go anyway.

他知道不方便,但他无论如何都想走。

50. He insists that it doesn't make any difference to him.

他坚持说这对他没有任何影响。

Specialized Sentences

1. A project execution is usually divided into some elementary phases, such as: engineering planning, procurement and transportation, and field construction.

一个工程项目的实施通常可分为几个基本阶段,例如:工程设计、采购和运输以及现场施工。

2. The contract number of this project is CJC78-8.

这个项目的合同号是 CJC78-8。

3. The seller (vendor) is Toyo Engineering Corporation (TEC) of Japan.

卖方(卖主)是日本的东洋工程公司(简称 TEC)。

4. The buyer (customer, client) is China National Technical Import Corporation (CNTIC).

买方(主顾、顾客)是中国技术进口总公司(简称 CNTIC)。

5. China National Chemical Construction Corporation (CNCCC) contracts for domestic and overseas chemical projects.

中国化工建设总公司(简称 CNCCC)承包国内和海外的化工工程。

6. Are you the seller's representative on the job site?

你是卖方的现场代表吗?

7. I am the buyer's general representative (GR).

我是买方的总代表(简称 GR)。

8. It is an inquiry (commercial and technical proposal, approval, agreement, protocol, annex, technical appendix) about this project.

这是这个项目的询价书(商务和技术报价书、批准书、协议、会议记录、附加条件、技术附件)。

9. There is much information in the technical proposal, which including: process flow, process description, capacity of the plant, performance of the product.

技术报价书中有很多资料,包括工艺流程、工艺说明、生产能力、产品特性。

10. The project team normally consists of project engineer, design engineer, schedule engineer, and various specialists.

项目工作组通常包括项目工程师、设计工程师、进度工程师以及各类专家。

11. We can evaluate the results of field construction by four criteria, which are quality, time, cost and safety.

我们可以通过四个指标来评价现场施工的效果,即质量、时间进度、费用和安全。

12. I am responsible for the technical (scheduling, inspection, quality control) work of this project (area).

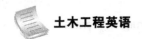

我负责这个项目(区域)的技术(进度、检查、质量控制)工作。

13. Would you tell us the technical characteristic about this project?

你能告诉我们有关这个工程项目的技术特性吗?

14. Please give a description about this project.

请对这个工程项目作一个叙述说明。

15. Do you have any reference materials about this project?

你有关于这个工程项目的参考资料吗?

16. We should work according to the overall schedule chart (the construction time schedule) of the project.

我们应该按照工程项目的总进度表(建设进度表)工作。

17. The effective date of this contract will begin from Dec. 30th, 2007.

这个合同的有效期将从2007年12月30日开始。

18. The seller will provide preliminary (final) technical documents for buyer in May.

卖方将于5月份向买方提供初步(最终)技术文件。

19. The basic (detailed) process design will be issued before August.

初步的(详细的)工艺设计资料将于8月前发出。

20. Our major planning items contain estimating of cost and construction schedule.

我们主要的计划内容包括费用预算和施工进度。

21. We shall have a design collecting (preliminary design, final design) meeting next month.

下个月我们将召开设计数据收集(初步设计、最终设计)会议。

22. Field erection work (civil work) will begin in October this year and complete on June 1st next year.

现场安装工作(土建工作)将自今年10月开始至明年6月1日完工。

23. The date of acceptance of this plant will be April 6th, 2008.

这座工厂的竣工日期拟订为2008年4月6日。

24. The seller's operating group (A crew of specialists) will remain on the job until guarantees are met.

卖方操作团队(专业团队)将在现场,一直工作到生产符合条件。

25. We must take the plant through the test run and finally into commercial operation.

我们必须使工厂通过试运转并最终投入商业运营。

26. Every month we shall establish construction schedule.

每个月我们都要制订建设进度计划。

27. We shall also make the project schedule report every day.

我们也将每天提出项目进度报告。

28. We have to change our plan for lack of materials (construction machinery, erection tools)

因缺少材料(施工机械、安装工具),我们只能改变计划。

29. What is your suggestion about this schedule?

你对这个计划进度有何建议？

30. Give me your opinion on this plan.

　　请把你对这个计划的意见告诉我。

31. We completed this task according to the drawing number SD-76.

　　我们按照图号 SD-76 的图纸完成了这项工作。

32. According to the technical standard (norm, rules of operation), the erection (alignment, testing) work is now getting on.

　　安装(校准、试验)工作正在根据技术标准(规范、操作规程)进行。

33. This is a plot plan (general layout, general arrangement, detail, section, erection, flow sheet, PID, assembly, civil, electrical, control and instrumentation, projection, piping, isometric) drawing.

　　这是一张平面(总平面、总布置、细部、剖面、安装、流程、带仪表控制点的管道、装配、土建、电气、控制和仪表、投影、管道、三维)图。

34. That is a general (front, rear, side, left, right, top, vertical, bottom, elevation, auxiliary, cut-away, birds eye) view.

　　那是全视(前视、后视、侧视、左视、右视、顶视、俯视、底部、立面、辅助、内部剖视、鸟瞰)图。

35. The general contractor must have an adequate amount of money due and payable.

　　总承包商必须有足够的到期应付款项。

36. Other provisions may include a definition of the rights of the parties on a default or termination of the contract.

　　其他规定还应包括在出现违约或合同终止情况下,对双方享有权利的规定。

37. The owner normally pays lump-sum prices which are guarded to varying degrees doesn't exceed maximum anticipated costs.

　　业主支付的一次性价格,应在不同程度上保证不超过最大预期成本。

38. The contractor is paid on completion of the contract a certain percentage of the amount.

　　承包商在合同完工的不同阶段将获得一定比例的合同款。

39. Is this a revised edition?

　　这是修订版吗?

40. Will it to be revised yet?

　　这张图还要修订吗?

41. Are there some modifications (revisions) on the drawing?

　　这张图上有修改(修正)吗?

42. The information to be placed in each title block of a drawing include drawing number, drawing size, scale.

　　每张图纸的图标栏应当包括图号、图纸尺寸、比例。

43. There are various types of lines on the drawing such as: border lines, visible lines, invisible lines, break lines, phantom lines.

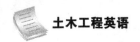

图上有各种形式的线条,诸如:边框线、实线、虚线、剖面线、辅助线。

44. We haven't received this drawing (instruction book, operation manual), please help us to get it.

我们还未收到这张图纸(说明书、操作手册),请帮助我们取得。

45. Please send us further information about this item.

请将有关这个项目的进一步的资料送交我们。

46. I want additional information on this.

我需要这方面的补充资料。

47. Please explain the meaning of this abbreviation (mark, symbol) on the drawing.

请解释图上这个缩写(标记、符号)的意义。

48. We comply with and carry out the GB standard (ANSI, BS, AFNOR, JIS and DIN) in this project.

在这个工程中我们遵守并执行中华人民共和国国家标准(GB)(美国标准、英国标准、法国标准、日本标准以及德国标准)。

49. Our store officer is responsible for the warehousing and issuing of materials.

我们的仓库管理员负责保管和发放材料。

50. We use Scientific-management system for material shortage and its control.

我们应用科学管理体系处理材料短缺及其调节。

Task Two   Sample Dialogue

Directions: *In this section, you are going to read several times the following sample dialogue about the relevant topic. Please pay special attention to five C's (culture, context, coherence, cohesion and critique) in the sample dialogue and get ready for a smooth communication in the coming task.*

Mary: Speaking of contracts, recently I've read a book on construction management and I find the different types of construction contracts very confusing.

Sam: What do you mean by confusing? What's your problem? Maybe I can help you.

Mary: Great. You know, three types are of primary interest, namely, lump-sum, unit-price and cost-plus-a-fixed-fee contracts.

Sam: That's right. But depending on the particular project, contractor, owner and many other complicated factors, different types would be preferred.

Mary: That's what lost me. What indeed are there differences?

Sam: A lump sum contract is the traditional means of procuring construction and involves a single "lump sum" price for all the works being agreed before the works begin. It is generally considered a beneficial contract agreement if the work is well defined when tenders are sought and significant changes to requirements are unlikely. This means that the contractor is able to accurately price the works they are being asked to carry out.

The terms of lump-sum contract provides that the owner will pay to the contractor a specified sum of money for the completion of a project conforming to the plans and specifications, but of course at specified intervals.

Mary: It seems to me that contractors would be at high risk for any unforeseen site conditions. What if there's a sharp increase in building materials prices after the contract is signed?

Sam: You can say that again. That's why many contractors would prefer unit-price contract, which provides that owner will pay to the contractor a specified amount of money for each unit of work completed in a project.

Mary: I see. How about CPFF contract? I remember under the terms of this contract, the owner agrees to reimburse the contractor for specified on-site costs, plus an additional management fee.

Sam: The cost-plus-fee contract is also referred to by the abbreviation of CPFF and represents a variant of a cost reimbursable contract in which the buyer provides reimbursement to the selling party for the allowable costs that have been accrued by the seller in the commission of the service, the creation, manufacture, delivery of the product, or in any other performance of the contracted work. It looks fair enough for contractors, but the profit margins would be rather thin. It's much less risky for contractors, so owners usually won't pay a good management fee.

Mary: That's a good point. Well, thanks for such a clear explanation.

Sam: You're welcome. Would you like another drink?

Mary: Sure. It'll be my treat.

Task Three   Simulation and Reproduction

Directions: *The class will be divided into three major groups, each of which will be assigned a topic. In each group, some students may be the teacher, while others may be students. In the process of discussion, please observe the principles of cooperation, politeness and choice of words. One of the groups will be chosen to demonstrate the discussion to the class.*

1) The performance of contract construction in practice.
2) An example of cost change of certain item in contract.
3) The importance of complying with contract.

Task Four   Discussion and Debate

Directions: *The class will be divided into two groups. Please choose your stand in regard to the following controversy and support your opinions with scientific evidences. Please refer to the specialized terms and classical sentences in the previous parts of this unit.*

Any construction project involves risks and there is no possibility to completely eliminate all

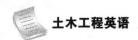 土木工程英语

the risks associated with a specific project. However, in an owner-contractor relationship at least, a common aim of owners appears to be avoid risks as far as possible by allocating as many risks as they can to the contractor. Please do some research and clarify what risks are most possible in a project for the owner?

V. After-class Exercises

1. *Match the English words in Column A with the Chinese meaning in Column B.*

A	B
1）general contractor	A）违约
2）agent	B）分包商
3）builder	C）合同
4）provision	D）总承包商
5）lump-sum	E）总金额
6）subcontractor	F）代理人
7）default	G）建筑商
8）termination	H）保证金
9）contract	I）规定（条款）
10）guaranteed cost	J）终止合同

2. *Fill in the following blanks with the words or phrases in the word bank. Change the forms if it's necessary.*

contract	machine	reimburse	board	warehousing
schedule	acceptance	contractor	preliminary	cement

1）I am responsible for the _____ work of this project.

2）The effective date of this _____ will begin from Dec. 30th, 2007.

3）The seller will provide _____ technical documents for buyer in May.

4）The owner agrees to _____ the contractor for specified on-site costs.

5）The above-mentioned _____ are widely used on the construction site.

6）There is a switch _____ mounted on the wall.

7）We have to change our plan for lack of construction _____.

8）Our store officer is responsible for the _____ and issuing of materials.

9）The date of _____ of this plant will be April 6th, 2008.

10）The general _____ must have an adequate amount of money due and payable.

3. *Translate the following sentences into English.*

1）这个合同的有效期将从 2007 年 12 月 30 日开始。

_____

_____

2）总承包商必须有足够的到期应付款项。

_____

_____

3）其他规定还应包括在出现违约或终止合同情况下,对双方享有权利的规定。

_____

_____

4）业主支付的一次性价格,应在不同程度上保证不超过最大预期成本。

_____

_____

5）承包商在合同完成时将获得一定比例的合同款。

_____

_____

4. *Please write an essay of about 120 words on the topic: Risks in Construction Contracts. Some specific examples will be highly appreciated and watch out the spelling of some specialized terms you have learnt in this unit.*

_____

_____

_____

_____

_____

_____

_____

_____

_____

_____

VI. Additional Reading

Engineering Design and Drawing

Engineering design is a systematic process by which a solution to the needs of humankind is obtained. The process is applied to problems (needs) of varying complexity. For example, mechanical engineers will use the design process to find an effective, efficient method to convert reciprocating motion to circular motion for the drive train in an internal combustion (燃烧) engine; electrical engineers will use the process to design electrical generating systems using

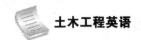

falling water as the power source; and materials engineers use the process to design ablative materials which enable astronauts to safely reenter the earth's atmosphere.

The vast majority of complex problems in today's high technology society depend on a solution not on single engineering discipline, but on teams of engineers, scientists, environmentalists, economists, sociologists, and legal personnel. Solutions are not only dependent regulations and political influence. As engineers, we are empowered with the technical expertise to develop new and improved products and systems, but at the same time, we must be increasingly aware of the impact of our actions on society and the environment in general and work conscientiously(认真地) toward the best solution in view of all relevant factors.

Design is the culmination of the engineering educational process; it is the salient feature that distinguishes engineering design is found in the curriculum guidelines of the Accreditation Board for Engineering and Technology (ABET). ABET accredits curricula in engineering schools and derives its membership from the various engineering professional societies. Each accredited curriculum has a well-defined design component that falls within the ABET statement on design read as follows.

Engineering design is the process of devising a system, component, or process to meet desired needs. It is the decision-making process, in which the basic sciences, mathematics and engineering sciences are applied to convert resources optimally to meet stated objectives and criteria, synthesis, analysis, construction, testing and evaluation. The engineering design component of a curriculum must include most of the following features: development of student creativity, use of open-ended problem statement and use of modern design theory and methodology(方法学), formulation of design problem statement and specification, consideration of alternative solutions, feasibility considerations, production process, concurrent engineering design, and detailed system descriptions. Further, it is essential to include a variety of realistic constraints such as economic factors, safety, reliability, ethics,and social impact.

Drafting or technical drawing is the means by which mechanical engineers design products and create instructions for manufacturing parts. A technical drawing can be a computer model or hand-drawn schematic showing all the dimensions necessary to manufacture a part, as well as assembly notes, a list of required materials and other pertinent(有关的) information. A U.S. mechanical engineer or skilled worker who creates technical drawings may be referred to as a drafter or draftsman. Drafting has historically been a two-dimensional process, but computer-aided design (CAD) programs now allow the designer to create in three dimensions. Instructions for manufacturing a part must be fed to the necessary machinery, either manually, through programmed instructions, or through the use of a computer-aided manufacturing (CAM) or combined CAD/CAM program. Optionally, an engineer may also manually manufacture a part using the technical drawings, but this is becoming an increasing rarity, with the advent of computer numerically controlled (CNC) manufacturing. Engineers primarily manually manufacture parts in the areas of applied spray coatings, finishes and other processes that cannot economically or practically be done by a machine. Working drawings are the complete set of standardized drawings specifying the manufacture and assembly of a product based on its design. The complexity of the design determines the number and types of drawings. Working drawings may be on more than one sheet and may contain written instructions called specifications. Working drawings are blueprints used for manufacturing products. Therefore, the set of drawings must: (1) completely describe the parts, both visually and dimensionally; (2) show the parts in assembly; (3) identify all the parts; (4) specify standard parts. The graphics and text information must be sufficiently complete and accurate to manufacture and assemble the product without error.

Generally, a complete set of working drawings for an assembly includes:

(1) Detail drawings of each nonstandard part.

(2) An assembly or subassembly drawing showing all the standard and nonstandard parts in a single drawing.

(3) A bill of materials (BOM).

(4) A title block.

A detail drawing is a dimensioned, multi-view drawing of a single part, describing the part's shape, size, material and surface roughness, in sufficient detail for the part to be manufactured based on the drawing alone. Detail drawings are produced from design sketches or extracted from 3-D computer models. They adhere (遵循) strictly to ANSI standards and the standard for the specific company, for dimensioning, assigning part numbers, notes, etc.

In an assembly, standard parts such as threaded fasteners and bearings are not drawn as details, but are shown in the assembly view. Standard parts are not drawn as details because they are normally purchased, not manufactured, for assembly.

For large assemblies or assembled with large parts, details are drawn on multiple sheets, and a separate sheet is used for the assembly view. If the assembly is simple or the parts are small, detail drawings for each part of an assembly can be placed on a single sheet.

Multiple details on a sheet are usually drawn at the same scale. If different scales are used, they are clearly marked under each detail. Also, when more than one detail is placed on a sheet, the spacing between details is carefully planned, including leaving sufficient room for dimensions and notes.

1. *Read the passage quickly by using the skills of skimming and scanning. And choose the best answer to the following questions.*

   1) What's the main meaning of the passage? _____.

      A. Engineering design is a systematic process

      B. Materials engineer's main work

      C. Mechanical engineer's good idea

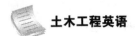

D. Electrical engineer' main duty

2) In the author's opinion, the design process is very complex and should be taken into account many factors, from the passage who should NOT join into the design team? _____.

　　A. Environmentalists　　　　　　　　B. Sociologists

　　C. Engineers　　　　　　　　　　　　D. Government officials

3) From the passage, we can know the meaning of words "ABET" (third paragraph) is _____.

　　A. a set of accredits standard

　　B. a kind of design method

　　C. a department of US government which responsible for engineering design

　　D. a set of law

4) It can be concluded from the passage that in the exercise and training of student, which character of the following is NOT included in the curriculum: _____.

　　A. Creativity　　　　　　　　　　　　B. New material

　　C. Ethics　　　　　　　　　　　　　　D. Economic factors

5) The title of the passage is _____.

　　A. the roles of engineers in manufacturing

　　B. the importance of mechanical design

　　C. engineering design

　　D. the process of machine design

6) Based on the passage, which of the follow is wrong? _____.

　　A. Standard parts needn't drawn as details.

　　B. For simple parts, sometimes they needn't to draw the detail working drawing.

　　C. For large assemblies details may be drawn on multiple sheets.

　　D. If different scales are used in a single sheet, they should be clearly marked under each detail.

7) Working drawings for an assembly MAY NOT includes _____.

　　A. BOM　　　　　　　　　　　　　　　B. title block

　　C. an assembly or subassembly drawing　　D. detail drawings of all parts

8) For detail drawings, Which one of the following statements is not true? _____.

　　A. Detail drawings should not have different scales

　　B. Detail drawings can be produced from design sketches or extracted from 3-D computer models

　　C. Detail drawings are dimensioned, multi-view drawing of a single part

　　D. Detail drawing should adhere strictly to ANSI standards

9) From the passage we can infer that ANSI is _____?

　　A. a set of law　　　　　　　　　　　B. a set of notification

C. an organization   D. IEEE government

10) The topic of the passage is _____.

      A. how to design working drawings

      B. working drawing's character and the key influence of how to draw a working drawing

      C. in the design process, what should be considered

      D. what is working drawing

2. *In this part, the students are required to make an oral presentation on either of the following topics.*

1) The features of Engineering design.

2) A complete set of working drawings for an assembly includes four parts, can you retell them briefly?

习题答案

Unit Five The Statue of Liberty

I. Pre-class Activity

Directions: *Please read the general introduction about* **The Statue of Liberty** *and tell something more about the statue to your classmates.*

The Statue of Liberty

The Statue of Liberty (Liberty Enlightening the World; French: La Liberté éclairant le monde) is a colossal neoclassical sculpture on Liberty Island in New York Harbor in New York City, in the United States. The copper statue, a gift from the people of France to the people of the United States, was designed by French sculptor Frédéric Auguste Bartholdi and built by Gustave Eiffel. The statue was dedicated on October 28, 1886.

The Statue of Liberty is a figure of Libertas, a robed Roman liberty goddess. She holds a torch above her head with her right hand, and in her left hand carries a tabula inscribed in Roman numerals with "JULY IV MDCCLXXVI" (July 4, 1776), the date of the U.S. Declaration of Independence. A broken chain lies at her feet as she walks forward. The statue became an icon of freedom and of the United States, a national park tourism destination, and is a welcoming sight to immigrants arriving from abroad.

Bartholdi was inspired by a French law professor and politician, Édouard René de Laboulaye, who is said to have commented in 1865 that any monument raised to celebrate the independence of USA would properly be a joint project of the French and U.S. peoples. Because of the post-war instability in France, work on the statue did not commence until the early 1870s. In 1875, Laboulaye proposed that the French finance the statue and the U.S. provide the site and build the pedestal. Bartholdi completed the head and the torch-bearing arm before the statue was fully designed, and these pieces were exhibited for publicity at international expositions.

The torch-bearing arm was displayed at the Centennial Exposition in Philadelphia in 1876, and in Madison Square Park in Manhattan from 1876 to 1882. Fundraising proved difficult, especially for the Americans, and by 1885 work on the pedestal was threatened by lack of funds. Publisher Joseph Pulitzer, of the New York World, started a drive for donations to finish the project and attracted more than 120,000 contributors, most of whom gave less than a dollar. The statue was built in France, shipped overseas in crates, and assembled on the completed pedestal on what was then called Bedloe's Island. The statue's completion was marked by New York's first ticker-tape parade and a dedication ceremony presided over by President Grover Cleveland.

The statue was administered by the United States Lighthouse Board until 1901 and then by the Department of War; since 1933 it has been maintained by the National Park Service. Public access to the balcony around the torch has been barred for safety since 1916.

II. Specialized Terms

Directions: *Please memorize the following specialized terms before the class so that you will be able to better cope with the coming tasks.*

accumulated damage 累积损伤
alternating load 交变载荷
alternating stress 交变应力
amorphous materials 非晶态材料
anchor v.使固定
assumption n. 假设
authority n. 权威人士
award a contract 授予合同
bind v.结合
binder n. 黏合料
bit n. 钻头
bitumen n. 沥青
bitumen coating 沥青外搪层，沥青外衬
bitumen felt 沥青纸
bitumen lining 沥青内搪层，沥青衬里
bituminous concrete 沥青混凝土
bituminous macadam 沥青碎石

bituminous waterproof membrane 沥青防水膜
blade n. 剪刀,叶片
blank flange 盲板法兰,盲板凸缘,管口盖板
blanking plate 封板
blast-furnace slag cement 炉渣水泥
blast-furnace 鼓风炉
blunt adj. 钝的
brittle damage 脆性损伤
brittle fracture 脆性断裂
buckle v.扣住，弯曲
chief engine 总工程师
constitutive relationship 本构关系
construction company 施工公司
construction crew 施工人员(总称)
construction material 施工材料

construction plan 施工计划

construction project 建筑项目,工程项目

consultant n. 咨询顾问

corrosion fatigue 腐蚀疲劳

course of bricks 砖层

crack gage 裂纹片

crack propagation 裂纹传播

creep deformation 徐变

creep fatigue 蠕变疲劳

crystal structure 晶体结构

cyclic hardening 循环硬化

damage and fracture 损伤与断裂

damage criterion 损伤准则

damage evolution equation 损伤演化方程

damage softening 损伤软化

damage strengthening 损伤强化

damage tensor 损伤张量

damage threshold 损伤阈值

damage variable 损伤变量

damage vector 损伤矢量

damage zone 损伤区

deck n. 甲板,舱面

deviatoric adj. 偏量的

ductile damage 延性损伤

ductile fracture 韧性断裂

environmental effect 环境效应

equation of strain compatibility 应变协调
方程

estimate v.估计,估价

excavation n. 挖土,掘土

fatigue n. 疲劳

fatigue crack 疲劳裂纹

fatigue damage 疲劳损伤

fatigue failure 疲劳失效

fatigue fracture 疲劳断裂

fatigue life 疲劳寿命

fatigue rupture 疲劳破坏

fatigue strength 疲劳强度

fatigue striations 疲劳条纹

fatigue threshold 疲劳阈值

feasible adj. 可行的,可能的

foundations n. 地基

fracture toughness 断裂韧性

homogeneous state of strain 均匀应变状态

hydroelectric project 水电项目

inhibit v.禁止, 抑制

initiate vt. 开始;发起;创始

isotropic adj. 各向同性

jammed adj. 挤满的,塞满的,拥挤不堪的

laminate vt. 碾压

linear elastic fracture mechanics 线弹性
力学

longitudinal adj. 纵向的

low cycle fatigue 应变疲劳

macroscopic damage 宏观损伤

microscopic damage 微观损伤

overloading effect 过载效应

pine n. 松木

plan n. 设计图

plastic deformation 塑性变形

pre-stress vt. 给……预加应力

random fatigue 随机疲劳

reinforce vt. 加固

resistance n. 抵抗力

rigid-beam 钢架桥

safe life 安全寿命

scaffold n. 脚手架

scale n. 比例尺

stress fatigue 应力疲劳

stress invariant 应力不变量

III. Watching and Listening

Task One The Statue of Liberty (I)

视频链接及文本

New Words

symbol n. 符号

signify vt. 表示,预示

ream n. 铰;令(500 张纸)

legislation n. 法律

pedestal n. 基架

concrete adj. 具体的

grueling adj. 折磨人的

cement n. 水泥,黏合剂

toss n. 投掷

enormous adj. 巨大的

iron n. 熨斗,铁

skeleton n. 骨架

trailing adj. 蔓生的

erect vt. 使竖立

Exercises

1. *Watch the video for the first time and answer the following questions.*

   1) How high is the statue? _____.

      A. 105 feet high   B. 115 feet high

      C. 150 feet high   D. 155 feet high

   2) When did people complete it? _____.

      A. A grueling autumn   B. A grueling winter

      C. A grueling spring   D. A grueling summer

   3) What do workers toss in for good luck? _____.

      A. Their own silver dollars   B. Their own clothes

      C. Their own shoes   D. Their own cigarettes

   4) What is Eiffel? _____.

      A. It's designed by Gustave Eiffel, who will build the famous Eiffel Tower in Paris

      B. It's designed by Gustave Eiffel, who will build the famous Triumphal Arch in Paris

      C. It's destroyed by Gustave Eiffel, who will build the famous Eiffel Tower in Paris

      D. It's destroyed by Gustave Eiffel, who will build the famous Triumphal Arch in Paris

   5) How many times is the sandal bigger than a human foot? _____.

      A. 22 times   B. 42 times

      C. 52 times   D. 32 times

2. *Watch the video again and decide whether the following statements are true or false.*

   1) Wrapping around the skeleton are 16,000 pounds of hand-sculpted copper. (　)

   2) In fact, maybe they do less than reams of legislation and paper and print. (　)

   3) I think a statue is not just a statue. I think symbols really matter. (　)

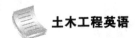

4) To hold a statue 150 feet high, the pedestal will be the biggest concrete structure in the world. (　　)

5) Over 200 men worked through a grueling winter to complete the statue. (　　)

3. *Watch the video for the third time and fill in the following blanks..*

Next, Liberty's _____ iron skeleton. It's _____ by Gustave Eiffel, who will build the famous Eiffel Tower in _____. The skeleton is 151 feet tall and with the pedestal, it's the _____ of a 30-_____ office block. Now for the outer layer. Wrapping around the skeleton are 60 000 pounds of hand-sculpted _____. This sandal is 32 times bigger than a _____ foot, _____ of the size of 879 _____. It's on the job trailing, often and 300 feet in erect. It's as difficult as it is _____.

4. *Share your opinions with your partners on the following topics for discussion.*

1) Can you tell us some stories about American history?

2) What is the most significant construction in America? Why? Please list your reasons.

Task Two　The Statue of Liberty (II)

New Words

framework n. 框架

rivets n. 铆钉

shell n. 壳，炮弹

robe n. 长袍

outstretch vt. 拉长

hazardous adj. 危险的

fatalities n. 死亡

winch n. 绞车

Mount Rushmore 拉什莫尔山

sculptor n. 雕刻家

acidize vt. 酸处理(酸化)

alliance n. 联盟

colonies n. 殖民地

beacon n. 烽火

huddled n. 杂乱一团

wretched adj. 可怜的

teeming adj. 充满的

视频链接及文本

Exercises

1. *Watch the video for the first time and answer the following questions.*

1) How many pieces of copper shell need to be fixed to the framework? _____.

    A. 200 pieces of copper shell　　　　B. 300 pieces of copper shell

    C. 400 pieces of copper shell　　　　D. 500 pieces of copper shell

2) How much does a finger nail weigh? _____.

    A. 2 and a half pounds　　　　B. 3 and a half pounds

    C. 4 and a half pounds　　　　D. 5 and a half pounds

3) Who did the sculptor Frederic Auguste Bartholdi model the face on? _____.

    A. His own grandmother　　　　B. His own grandfather

    C. His own mother　　　　D. His own father

4) What is the statue's official name? _____.

    A. American Liberty   B. Liberty of America

    C. America of the world   D. Liberty enlightening the world

5) Where is the Statue of Liberty? _____.

    A. At the entries to New Jersey Harbor   B. At the entries to New York Harbor

    C. At the front of New York Harbor   D. At the front of New Jersey Harbor

2. *Watch the video again and decide whether the following statements are true or false.*

    1) The statue's outstretched arm is 42 feet long. (　)

    2) A finger nail weighs 3 and a half pounds.(　)

    3) The scale of Liberty is imaginable. (　)

    4) It takes 25 years for Liberty to acidize and turn blue. (　)

    5) A functioning lighthouse until 1902. (　)

3. *Watch the video for the third time and fill in the following blanks.*

At first, the symbol of the _____ and friendships between France and the 13 _____ in the American _____. It will come to _____ much more. At the entries to New York _____, the Statue of Liberty becomes a _____ to the world and a welcome to millions. Later, a poem by Emma Lazarus in her _____, celebrates America as a land of _____: "Give me your tired, your poor, your huddled masses, yearning to _____ free. The wretched refuse of your teeming shore. Send these, the _____, tempest to me, I lift my lamp beside the golden door!"

4. *Share your opinions with your partners on the following topic for discussion.*

After watching the video and knowing how the Statue of Liberty was built, will you think of the People's Heroes Monument majestically and austerely standing before the Tiananmen Square? Will you think of those heroes who lost their lives in the fight for the independence of the country and freedom of people? Please share your opinion with your classmates.

IV. Talking

Task One   Classical Sentences

Directions: *In this section, some popular sentences are supplied for you to read and to memorize. Then, you are required to simulate and produce your own sentences with reference to the structure.*

General Sentences

    1. I'm going shopping because I need to buy some clothes.

    我想去逛逛商店,因为我需要买一些衣服。

    2. Yesterday was such a beautiful day and we decided to go for a drive.

    昨天天气很好,所以我们开车出去玩了一趟。

3. What are you going to wear today?

你今天打算穿什么？

4. I'm going to wear my blue suit. Is that all right?

我要穿我的蓝色西装。怎么样？

5. I have some shirts to send to the laundry.

我有一些衬衫要送到洗衣房去。

6. You ought to have that coat cleaned.

你应该把那件外套洗一下。

7. I've got to get this shirt washed and ironed.

我得把这件衬衫洗了烫一烫。

8. All my suits are dirty. I don't have anything to wear.

我所有的西装都脏了。我没有穿的了。

9. You'd better wear a light jacket. It's chilly today.

你最好穿件薄夹克。今天很冷。

10. This dress doesn't fit me anymore.

这件衣服我穿已经不合身了。

11. These shoes are worn-out. They've lasted a long time.

鞋已经磨破了。我穿了好长时间。

12. Why don't you get dressed now? Put on your work clothes.

为什么还不换装？穿上你的工作服呀。

13. My brother came in, changed his clothes and went out again.

我哥哥进来了，换好衣服后又出门了。

14. I didn't notice you were wearing your new hat.

我都没注意到你戴了新帽子。

15. If you want a towel, look in the linen closet.

如果你想要毛巾，到放毛巾的壁橱里找。

16. My brother wants to learn how to dance.

我弟弟想学跳舞。

17. Which would you rather do, go dancing or go to cinema?

你想干什么？去跳舞还是去看电影？

18. I'd like to make an appointment to see Mr. Cooper.

我想约好时间去看看库珀先生。

19. Would you like to arrange for a personal interview?

你想安排一场个人采访吗？

20. Your appointment will be next Thursday at 10 o'clock a.m.

你的会面安排在下周四上午十点。

21. I can come any day except Thursday.

除了星期四，我每天都能来。

22. He wants to change his appointment from Monday to Wednesday.
他想将会面从周一改到周三。

23. She failed to call the office to cancel her appointment.
她没能打电话让办公室取消她的预约。

24. I'm going to call the employment agency for a job.
我要打电话给职业介绍所申请一份工作。

25. Please fill in this application form.
请填写这张申请表。

26. Are you looking for a permanent/temporary position?
你想应聘一个长期/临时职位吗?

27. I'm going to call a plumber to come this afternoon.
我打算今天下午叫个管道工来。

28. I couldn't keep the appointment because I was sick.
因为我病了,不能守约了。

29. Please call before you come, otherwise we might not be home.
请在你来之前打个电话,不然我们有可能不在家。

30. Will you please lock the door when you leave?
你离开时把门锁上,好吗?

31. I went to see my doctor for a check-up yesterday.
我昨天去我医生那儿检查了。

32. The doctor discovered that I'm a little overweight.
医生发现我有些超重。

33. He gave me a chest X-ray and took my blood pressure.
他让我做了个 X 光胸透,又给我量了量血压。

34. He told me to take these pills every four hours.
他叮嘱我,每四小时吃一次药。

35. Do you think the patient can be cured?
你觉得这个病人能治愈吗?

36. —What did the doctor say?
—The doctor advised me to get plenty of exercise.
—医生说什么?
—医生建议我多做些运动。

37. If I want to be healthy, I have to stop smoking.
如果我想健康的话,我得戒烟。

38. It's just a mosquito bite. There's nothing to worry about.
只是被蚊子叮了下,不必担心。

39. —How are you feeling today?
—Couldn't be better.

—今天感觉怎样？
—非常好。

40. I don't feel very well this morning.
今天早上我感到不舒服。

41. I was sick yesterday, but I'm better today.
昨天我病了,不过今天好些了。

42. My fever is gone, but I still have a cough.
我已经不发烧了,不过还在咳嗽。

43. —Which of your arms is sore?
—My right arm hurts. It hurts right here.
—你的哪个胳膊疼？
—我的右胳膊疼,就在这儿。

44. —What's the matter/wrong/the trouble with you?
—I've got a pain in my back.
—你哪里不舒服？
—我背疼。

45. —How did you break your leg?
—I slipped on the stairs and fell down. I broke my leg.
—你的腿怎么伤的？
—我在楼梯上摔了一跤,腿就受伤了。

46. Your right hand is swollen. Does it hurt?
你的右手肿了,疼吗？

47. It's bleeding. You'd better go to see a doctor about that cut.
你的伤口在流血,最好去医生那看看。

48. Why do you dislike the medicine so much?
为什么你这么不喜欢药？

49. I didn't like the taste of the medicine, but I took it anyway.
我不喜欢药的味道,但是无论如何我都吃了。

50. I hope you'll be well soon.
希望你快些好。

Specialized Sentences

1. Civil engineering, the oldest of the engineering specialties, is the planning, design, construction, and management of the built environment.
土木工程,这个最古老的工程专业,是指对建筑的规划、设计、建造和现场管理。

2. This Civil engineering environment includes all structures built according to scientific principles, from irrigation and drainage systems to rocket-launching facilities.
土木工程环境包括按科学原理所建造的一切结构,从灌溉系统、排水系统到火箭发射设备。

3. Civil engineers build roads, bridges, tunnels, dams, harbors, power plants, water and sewage systems, hospitals, schools, mass transit, and other public facilities essential to modern society and large population concentrations.

土木工程师们修建道路、桥梁、隧道、水坝、港口、电厂、供水排水系统、医院、学校、公共交通设施和其他公共设施,这些对于建设现代化社会和人口集中居住地不可或缺。

4. Civil engineers also build privately owned facilities such as airports, railroads, pipelines, skyscrapers, and other large structures designed for industrial, commercial, or residential use.

土木工程师们也建设私营设施,例如机场、铁路、管网、摩天大楼和其他大型建筑物,它们被设计用于工业、商业和居住等各种用途。

5. Civil engineers plan, design, and build complete cities and towns, and more recently have been planning and designing space platforms to house self-contained communities.

土木工程师规划、设计和建设完整的城市和乡镇;近年来,他们已经开始规划和设计空间平台来构建自给自足型社区。

6. Because it is so broad, civil engineering is subdivided into a number of technical specialties. Depending on the type of project, the skills of many kinds of civil engineer specialists may be needed.

因为它的面太广,所以土木工程被分成许多技术专业。各专业的土木工程专家所需要的技能取决于工程项目的类型。

7. When a project begins, the site is surveyed and mapped by civil engineers who locate utility placement—water, sewer, and power lines.

当一个项目开始时,土木工程师会测绘现场,会确定给排水设施和电力线路的实际位置。

8. Geotechnical specialists perform soil experiments to determine if the earth can bear the weight of the project.

岩土工程专家们进行土壤实验来确定地基是否能承受工程项目的自重。

9. Environmental specialists study the project's impact on the local area: the potential for air and groundwater pollution, the project's impact on local animal and plant life, and how the project can be designed to meet government requirements aimed at protecting the environment.

环境专家们研究项目对当地的影响:例如潜在的空气和地下水资源污染可能;工程项目对当地动植物的影响;如何满足保护环境的法律要求并把工程项目设计好。

10. Transportation specialists determine what kind of facilities are needed to ease the burden on local roads and other transportation networks that will result from the completed project.

运输专家确定需要什么设施来减轻由项目带来的对当地道路和其他运输网络的

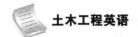

压力。

11. Structural specialists use preliminary data to make detailed designs, plans, and specifications for the project.

结构专家用初步数据为工程项目做出详细的设计、规划和说明。

12. Supervising and coordinating the work of these civil engineer specialists, from beginning to end of the project, are the construction management specialists.

从工程开工到结束整个过程,建筑管理专家都监督和协调着土木工程师的工作。

13. Based on information supplies by the other specialists, construction management civil engineers estimate quantities and costs of materials and labor, schedule all work, order materials and equipment for the job, hire contractors and subcontractors, and perform other supervisory work to ensure the project is completed on time and as specified.

基于其他专家所提供的信息,建筑管理土木工程师预估所需的材料和人工的数据,对所有的工作提出计划,为各工序订购材料和设备,签订承包合同和转包合同,完成其他监督工作以确保工程按期按质完成。

14. Throughout any given project, civil engineers make extensive use of computers. Computers are used to design the project's various elements (computer-aided design, or CAD) and to manage it.

对于任何一个给定的工程项目,土木工程师始终都要充分利用计算机。用计算机来设计工程的各要素(计算机辅助设计,CAD)以及管理这个工程项目。

15. Computers are necessity for the modern civil engineer because they permit the engineer to efficiently handle the large quantities of data needed in determining the best way to construct a project.

对于现代土木工程师而言,计算机是必备的工具,因为它们可以使工程师高效地处理大量的用来制定工程最佳施工方法的数据。

16. Using computers, structural engineers determine the forces a structure must resist: its own weight, wind and hurricane forces, temperature changes that expand or contract construction materials, and earthquakes.

利用计算机,结构工程师计算一个结构所承受的各种力:建筑物自重、风力和飓风压力、温度变化而引起的建筑材料的热胀冷缩的应力和地震力。

17. The Water resources engineering projects help prevent floods, supply water for cities and for irrigation, manage and control rivers and water runoff, and maintain beaches and other waterfront facilities.

水资源工程项目可以帮助防洪,为城市供水和灌溉,管理和控制河流和排水,维护海滩和其他沿岸设施。

18. Civil engineers who specialize in Geo-technical Engineering analyze the properties of soils and rocks that support structures and affect structural behavior.

从事岩土工程领域工作的土木工程师分析支撑建筑物并影响结构特性的土壤和岩石的性质。

19. They evaluate and work to minimize the potential settlement of buildings and other structures that stems from the pressure of their weight on the earth.

他们计算建筑和其他结构由于自重压力可能引起的沉降,并采取措施使之减少到最小。

20. These engineers also evaluate and determine how to strengthen the stability of slopes and fills and how to protect structures against earthquakes and the effects of groundwater.

这些工程师也评估和确定如何加强边坡和充填物的稳定性,如何保护建筑物使其免受地震和地下水的影响。

21. In environmental engineering, civil engineers design, build and supervise systems to provide safe drinking water and to prevent and control pollution of water supplies, both on the surface and underground.

在环境工程中,土木工程师设计、建造和监督提供安全饮用水的系统,以及防止和控制地表水以及地下水资源污染的系统。

22. In transportation engineering, civil engineers working in this specialty build facilities to ensure safe and efficient movement of both people and goods.

从事运输工程专业的土木工程师通过建设公用设施保证人和物的安全有效流动。

23. Hydraulic pump is the power unit of the hydraulic puller (hydraulic press, hydraulic pipe bender, hydraulic jack).

油压泵是油压拉出器(油压机、油压弯管机、油压千斤顶)的动力装置。

24. A welder's kit contains electrode holder, welding torch, helmet shield, portable electrode heating box and temperature measuring pen.

一名焊工的成套工具包括焊钳、焊炬、面罩、手提式焊条加热箱和测温笔。

25. The diameter of this wire rope (hemp rope, sling) is three-fourth inches.

这钢丝绳(麻绳、吊索)的直径为 3/4 英寸。

26. The lifting capacity of this chain hoist (hydraulic jack, screw jack) is 5 tons.

这个吊链(油压千斤顶、螺旋千斤顶)的起重能力为 5 吨。

27. The vise (parallel-jaw vice) is available to all of the bench work.

所有的钳工工作都可使用台钳(平口钳)。

28. Grease gun and oiler are the lubrication service tools for machinery.

油枪和注油器都是机械润滑维护工具。

29. Torque wrenches offer the precision measurement needed to tighten fasteners.

力矩扳手可以提供紧固螺栓所需的精确力矩计量。

30. The measuring unit of torque wrench is pound-inch or kilogram-centimeter.

力矩扳手的计量单位为磅/英尺或者千克/厘米。

31. Is the machine accompanied with some tools (spare parts, accessories)?

这台机器随机带有一些工具(备件、附件)吗?

32. Shall we use a special tool for this job?

我们干这活要使用专用工具吗?

33. Could you tell us how to use (operate, repair, maintain, clean, adjust) this new tool?

你能告诉我们如何使用(操作、修理、维护、清理、调整)这个新工具吗？

34. The tool gets out of order and we must remedy its trouble.

这工具有毛病,我们必须排除它的故障。

35. The tool is out of repair, it needs an overhaul.

这工具失修,需要拆修。

36. From your explanation I shall easily handle it.

听了你的说明,我将更轻松地使用它。

37. The tools must be well kept.

工具必须妥善保管。

38. There are some material warehouses on the construction site.

在工地上有一些材料仓库。

39. Our store officer is responsible for the warehousing and issuing of materials.

我们的仓库管理员负责保管和发放材料。

40. We use Scientific-management system for material shortage and its control.

我们应用科学管理体系处理材料短缺及其调节。

41. These materials are imported from abroad.

这些材料是从国外进口的。

42. What is the feature about it?

这些材料的特性是什么？

43. The construction material answers our purpose satisfactorily.

这种建筑材料能满足我们的需要。

44. The average traffic gasoline consumption of this lorry is 0.3 liter per kilometer (l/km).

这台货车的平均行车油耗为每千米 0.3 升(升/千米)。

45. Hydraulic oil (lubrication oil) which having a viscosity of about 4.5°E at 50℃ can be used for this vehicle (machine).

具有恩氏黏度 4.5°E(50℃)的液压油(润滑油)可用于此车辆(机器)。

46. This special oil comes from the "SHELL" company (CALTEX, MOCBIL, GULF, ESSO, CASTROL, BP).

这种特种油来自"壳牌"公司(加德士、飞马、海湾、埃索、卡斯特罗、英国石油公司)。

47. Cement steel and timber are the most important construction materials used in civil engineering.

钢材和木材是土建工程中最重要的建筑材料。

48. Typical structural steel shapes include beams, channels, angles and tees.

典型的结构型钢包括工字钢、槽钢、角钢和 T 字钢。

49. There are four broad classifications of steel: carbon steels, alloy steels, high-strength low-alloy steels and stainless steels.

钢材大致可分为四类,即碳素钢、合金钢、高强度低合金钢和不锈钢。

50. Copper, zinc, lead, aluminum, bronze and brass are nonferrous metals or alloys.

铜、锌、铅、铝、青铜和黄铜都是有色金属或合金。

Task Two   Sample Dialogue

Directions: *In this section, you are going to read several times the following sample dialogue about the relevant topic. Please pay special attention to five C's (culture, context, coherence, cohesion and critique) in the sample dialogue and get ready for a smooth communication in the coming task.*

A Job Interview

(*Several equipment operators and dump truck drivers are needed for the Hydro project. After the job vacancies have been advertised in the newspaper, the job seekers come to the site office for interviews. Mr. Zhang is receiving them.*)

Job Seeker: Good morning, I am asking about your advertisement for operators.

Zhang: What kind of job are you looking for?

Job Seeker: I am a bulldozer operator. So I want to work here as a bulldozer operator.

Zhang: I am sorry. The vacancies for bulldozer operators have been filled. You should have tried earlier.

Job Seeker: What a pity!

Zhang: If you like, you can fill in this application form and leave us your wechat number. As soon as we have a job that suits you, we will give you a message.

Job Seeker: Sorry, I don't have a wechat. I have put my telephone number here on this form. Please call me once you have a job for me.

Zhang: Right. I hope we'll soon have a job for you.

Job Seeker: Thanks a lot.

＊＊   ＊＊   ＊＊   ＊＊   ＊＊

Job Seeker: Excuse me, I saw in your advertisement that you have some vacancies for dump truck drivers. Are they still available?

Zhang: We still have two vacancies for ten-wheel dump truck drivers.

Job Seeker: Lucky enough. I am looking for a job as a driver.

Zhang: May I see your driving license?

Job Seeker: Certainly, here it is.

Zhang: How long have you been driving this kind of truck?

Job Seeker: Ten years. Here is a letter of recommendation from my previous employer.

Zhang: Fine. May I ask why you quitted the job there?

Job Seeker: I live in San Ignacio. The company I worked in is too far from my home. Besides, the transportation is not convenient.

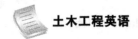

Zhang: Do you have your ID with you?

Job Seeker: Yes. Here you are.

Zhang: And your Social Security?

Job Seeker: I had one, but I lost it last week. If you accept me, I will apply to the Social Security Board for a new one. I think I can get it within a week.

Zhang: OK. This is an application form. Please fill it in. Tomorrow you come here for a driving test. If you pass it, you will be a probationary worker for two weeks. That means we will try you for two weeks. If you prove to be a qualified driver during that period, you will become a formal worker and can work with us until this Hydro project is completed.

Job Seeker: How much do you pay, please?

Zhang: The starting wage is 25 dollars for a day shift and 30 for a night shift. One shift is seven hours and a half, with extra pay for overtime work, if any. You will have the chance to get a rise if you prove to be a good worker.

Job Seeker: Do I have to live on the site camp? If not, how do I get to the job site every day?

Zhang: We provide free transportation for all our workers. You said you live in San Ignacio? Most of our local workers come from this area. We have a bus that carries our workers from our bus stop in San Ignacio directly to the job site.

Job Seeker: That's good. So what time do I come here for the test tomorrow?

Zhang: You can come on our bus which leaves San Ignacio at six thirty in the morning. When you arrive, please come to me. I'll arrange the test for you.

Job Seeker: All right. What if the bus driver won't let me get on the bus?

Zhang: Don't worry. Take this note and show it to the driver.

Job Seeker: Thank you, sir, I'll see you tomorrow.

Zhang: OK, see you tomorrow and good luck.

Task Three   Simulation and Reproduction

Directions: *The class will be divided into two major groups, each of which will be assigned a topic. In each group, some students may be the teacher, while others may be students. In the process of discussion, please observe the principles of cooperation, politeness and choice of words. One of the groups will be chosen to demonstrate the discussion to the class.*

1) Simulate a job interview.

2) Suppose you were rejected by a company, what would you do?

Task Four   Discussion and Debate

Directions: *The class will be divided into two groups. Please choose your stand in regard to the* *following controversy and support your opinions with scientific evidences. Please refer to the specialized terms and classical sentences in the previous parts of this unit.*

After learning Golden Gate Bridge and the Statue of Liberty, which one do you think is the symbol of America? Why? Please choose your side and give your reasons. Which civil-enginecring construction do you think can carry the spirit of China, the Great Wall, Temple of Heaven in Beijing, or the Forbidden City? Why?

Furthermore, how do you think of Chinese traditional culture featuring peace, harmony and cooperation in comparison with the western culture, which values individualism and competition stemming from the jungle law? Can you illustrate this cultural gap by citing the performances of the two countries, China and American, in dealing with the international hot spots?

V. After-class Exercises

1. *Match the English words in Column A with the Chinese meaning in Column B.*

A	B
1) project manager	A) 现场管理人员
2) site management	B) 施工计划
3) construction project	C) 总工程师
4) construction crew	D) 项目经理
5) technical data	E) 咨询顾问
6) specification	F) 建筑材料
7) construction plan	G) 建筑项目
8) consultant	H) 技术资料
9) chief engineer	I) 施工人员
10) construction material	J) 技术规范

2. *Fill in the following blanks with the words or phrases in the word bank. Change the forms if it's necessary.*

import	short	move	experiment	answer
explanation	machine	minimize	survey	consumption

1) When a project begins, the site is _____ and mapped by civil engineers who locate utility placement—water, sewer, and power lines.

2) Geotechnical specialists perform soil _____ to determine if the earth can bear the weight of the project.

3) They evaluate and work to _____ the potential settlement of buildings and other structures that stems from the pressure of their weight on the earth.

4) In transportation engineering, civil engineers working in this specialty build facilities to

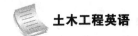

ensure safe and efficient _____ of both people and goods.

5）Grease gun and oiler are the lubrication service tools for _____.

6）We use Scientific-management system for material _____ and its control.

7）From your _____ I shall easily handle it.

8）These materials are _____ from abroad.

9）The construction material _____ our purpose satisfactorily.

10）The average traffic gasoline _____ of this lorry is 0.3 liter per kilometer.

3. *Translate the following sentences into English.*

1）这台工具随机带有一些工具(备件、附件)吗?

_____

2）这工具有毛病,我们必须排除它的故障。

_____

3）在工地上有一些材料仓库。

_____

4）钢材和木材是土建工程中最重要的建筑材料。

_____

5）钢材大致可分为四类,即碳素钢、合金钢、高强度低合金钢和不锈钢。

_____

4. *Please write an essay of about 120 words on the topic：The symbolic construction of China. Some specific examples will be highly appreciated and watch out the spelling of some specialized terms you have learnt in this unit.*

_____

VI. Additional Reading

The Statue of Liberty

[A] Out of all of America's symbols, none has proved more enduring or evocative(引起回忆的) than the Statue of Liberty. This giant figure, torch in hand and clutching a stone tablet, has for a century acted as a figurehead for the American Dream; indeed there is probably no more immediately recognizable profile in existence.

[B] It's worth remembering that the statue is for Americans at least—a potent reminder that the USA is a land of immigrants: it was New York Harbor where the first big waves of European immigrants arrived, their ships entering through the Verrazano Narrows to round the bend of the bay and catch a first glimpse of "Liberty Enlightening the World"—an end of their journey into the unknown, and the symbolic beginning of a new life.

[C] These days, although only the very wealthy can afford to arrive here by sea, and a would-be immigrant's first (and possibly last) view of the States is more likely to be the customs check at JFK Airport, Liberty remains a stirring(搅动) sight, with Emma Lazarus's poem, The New Colossus, written originally to raise funds for the statue's base, no less quotable than when it was written:

[D] Here at our sea-washed, sunset gates shall stand a mighty woman with a torch, whose flame is the imprisoned lightning, and her name Mother of Exiles. From her beacon-hand glows world-wide welcome; her mild eyes command the air-bridged harbor that twin cities frame. "Keep ancient lands, your storied pomp!" cries she, with silent lips. "Give me your tired, your poor, your huddled(拥挤的) masses yearning to breathe free, the wretched refuse to your teeming shore. Send these, the homeless, tempest-tost to me, I lift my lamp beside the golden door."

[E] The statue, which depicts Liberty throwing off her shackles and holding a beacon to light the world, was the creation of the French sculptor Frédéric Auguste Bartholdi, who crafted it a hundred years after the American Revolution in recognition of solidarity(齐心协力) between the French and American people (though it's fair to add that Bartholdi originally intended the statue for Alexandria in Egypt).

[F] Bartholdi built Liberty in Paris between 1874 and 1884, starting with a terracotta(赤陶土) model and enlarging it through four successive versions to its present size, construction of thin copper sheets bolted together and supported by an iron framework designed by Gustave Eiffel. The arm carrying the torch was exhibited in Madison Square Park for seven years, but the whole statue wasn't officially accepted on behalf of the American people until 1884, after which it was taken apart, crated up and shipped to New York.

[G] It was to be another two years before it could be properly unveiled: money had to be collected to fund the construction of the base, and for some reason Americans were unwilling—

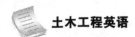

or unable—to dip into their pockets. Only through the campaigning efforts of newspaper magnate (权贵、富豪) Joseph Pulitzer, a keen supporter of the statue, did it all come together in the end.

[H] Richard Morris Hunt built a pedestal around the existing star-shaped Fort Wood and Liberty was formally dedicated by President Cleveland on October 28, 1886, in a flag-waving shindig that has never really stopped. The statue was closed for a few years in the mid-1980s for extensive renovation, in 1986, fifteen million people descended on Manhattan for the statue's centennial celebrations.

[I] Today you can climb steps up to the crown, but the cramped stairway though the torch sadly remains closed to the public. Don't be surprised if there's an hour-long wait to ascend. Even if there is, Liberty Park's views of the lower Manhattan skyline, the twin towers of the World Trade Center lording it over the jutting(突出) teeth of New York's financial quarter, are spectacular enough.

1. *Read the passage quickly by using the skills of skimming and scanning. And choose the best letter standing for each paragraph above in response to the following sentences.*

   1) Money had to be collected to fund the construction of the base.

   2) It was New York Harbor where the first big waves of European immigrants arrived.

   3) This giant figure, torch in hand and clutching a stone tablet, has for a century acted as a figurehead for the American Dream.

   4) The cramped stairway though the torch sadly remains closed to the public.

   5) Liberty was formally dedicated by President Cleveland on October 28, 1886.

   6) The arm carrying the torch was exhibited in Madison Square Park for seven years.

   7) A would-be immigrant's first view of the States is more likely to be the customs check at JFK Airport.

   8) "Keep ancient lands, your storied pomp!" cries she, with silent lips.

   1) _____   2) _____   3) _____   4) _____

   5) _____   6) _____   7) _____   8) _____

2. *In this part, the students are required to make an oral presentation on either of the following topics.*

   1) A brief introduction of the Statue of Liberty.

   2) A welcome speech for the people going to the America.

习题答案

Unit Six Project Cost Estimates

I. Pre-class Activity

Directions: *Please read the general introduction about* **Construction Estimators** *and tell something more to your classmates.*

Construction Estimators

Construction estimators analyze the costs of and prepare estimates on construction projects. They may specialize in estimating costs for civil engineering, architectural, structural, electrical and mechanical construction projects; or they may specialize in estimating costs for one construction trade in particular, such as electrical. They are employed by residential, commercial and industrial construction companies and major electrical, mechanical and trade contractors. In some cases they may be self-employed, and in smaller organizations estimators may also perform other tasks. As a Construction Estimator, your duties may include the following:

Prepare estimates of probable costs of materials, labor and equipment, subcontracts for construction projects based on contract bids, quotations, schematic drawings and specifications.

Advise on tendering procedures, examine and analyze tenders, recommend tender awards and conduct negotiations.

Set up cost monitoring and reporting systems and procedures.

Prepare cost and expenditure statements and forecasts at regular intervals for the duration of a project.

Prepare and maintain a directory of suppliers, contractors and subcontractors.

Liaise, consult and communicate with engineers, architects, owners, contractors and subcontractors, prepare economic feasibility studies on changes and adjustments to cost estimates.

Manage and co-ordinate construction projects, and prepare construction progress schedules.

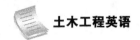

II. Specialized Terms

Directions: *Please memorize the following specialized terms before the class so that you will be able to better cope with the coming tasks.*

accelerating water reducer 速凝型减水剂

accelerator n. 速凝剂

air-entraining agent 引气剂

arising pipe 升浆管

backfill grouting 回填灌浆

backfilling of hole 封孔

blended adj. 混合的

bloating bond breaker 脱模剂

boundary n. 边界,分界线

bow saw 弓锯,弧锯

brad n. 角钉,平头钉

brittle material 脆性材料

brush n. 刷子

building permission 建筑开工许可证

clamp n. 卡子

climbing cone 爬升锥

cock n. 旋塞

collapse v. 倒塌

coloring admixture 着色剂

compression n. 压力

compressive strength 抗压强度

consolidation grouting 固结灌浆

contact grouting 接触灌浆

copper water stop 铜止水

curtain grouting 帐幕灌浆

dam bracket 大坝支架

deflect v. 偏斜

depression n. 坑洼

distance piece 横支架

distribution bar 分布钢筋

double-curvature formwork 双曲模板

dowel n. 插筋

early-strength admixture 早强剂

expansion agent 膨胀剂

final strength 终凝强度

flat formwork 平面模板

form erecting formwork 立模

formwork shuttering 模板

impose v. 将……强加于

postulate v. 假定

precedent n. 先例

principle of virtual work 虚功原理

probability n. 概率

proceed vi. 继续进行

processor n. 抛射物,发射体

prompt vt. 促使,推动,提示

prompt line 提示行

property n. 财产,地产

purpose n. 目的,用途

rational adj. 合理的

restore v. 恢复,修复

restrain v. 抑制,遏止

resume v. 重新开始,继续

retarding agent 抑制剂

rust n. 锈,铁锈

sack n. 麻袋,纸袋,塑料袋

sacred adj. 神圣的

sag v. 下垂

scissors n. 剪刀,剪子

scraper n. 三角刮刀

segment n. 部分;切片,断片

set square 三角板

shovel n. 铲子,铁锹

site management 现场管理人员(总称),现

场管理机构

sketch n. 草图

skyscraper n. 摩天大楼

slab n. 厚片，厚块

sliding n. 滑，移动

spade n. 锹，铲

single crystalline materials 单晶体材料

stem vi. 起源于

strain ellipsoid 应变椭球

strain fatigue 应变疲劳

strain invariant 应变不变量

stream v. 流，流动，流出

strength criterion 强度判据，强度准则

stress amplitude 应力幅值

stress cycle 应力循环

stress fatigue 应力疲劳

stress intensity factor 应力强度因子

stress invariant 应力不变量

stress ratio 应力比

string vt. 给……装弦

strung adj. 捆扎的，串起的，紧张的

subsoil n. 底土，下层土，天然地基

sun-baked 日晒的，晒干的

suspend v. 悬，挂，吊

tensile adj. 可拉长的，张力的

tension n. 拉力

theory of elasticity 弹性理论

thrust n. 推力，刺，力推

to lay the foundations 打地基

to prefabricate 预制

torque n. 扭转力矩

transverse adj. 横向的

ultimate adj. 终极的

vault and invert formwork 顶拱及底拱模板

vertical bar 竖筋

water seepage 渗水剂

winter protection 冬季保温

III. Watching and Listening

Task One　The Construction Estimator

视频链接及文本

New Words

surveyor n. 测量员

certified adj. 被鉴定的

quantity n. 数量

professional adj. 专业的

Exercises

1. *Watch the video for the first time and choose the best answers to the following questions.*

　　1) The following jobs should be done by construction estimators except _____.

　　　　A. finding a job to bid　　　　　　B. analyzing all the risks

　　　　C. reading the specifications　　　D. bargaining with the clients

　　2) Why did I find it hard to work as a construction estimator? _____.

　　　　A. Because not every building can be built the same way

　　　　B. Because not every structure is going to be on schedule or meet the schedule desires

　　　　C. Because inexperience made it very difficult to find work in that field

　　　　D. Because a construction estimator is not just about reading specifications, but putting

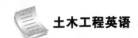
the project in their mind

3) Why did they change the opinion of me? _____.

    A. I progressed into a professional quantity surveyor

    B. I became a member of Keynes

    C. I gained a lot of certificates

    D. I became an expert in construction estimating

4) I will build the other community for _____.

    A. my kids

    B. my father

    C. myself

    D. everybody who wants to become a construction estimator

5) This passage is writing to _____.

    A. teach people how to become a construction estimator

    B. introduce the occupation of construction estimator

    C. show the advantage of becoming a construction estimator

    D. show the disadvantage of becoming a construction estimator

2. *Watch the video again and decide whether the following statements are true or false.*

1) The job of a construction engineer is to find a job to bid on. (　)

2) An architectural estimate is just a reading specification. (　)

3) Understanding how buildings form is not just a matter of cost. (　)

4) All buildings will be completed on time or meet the requirements. (　)

5) When the speaker graduated, he found it difficult to get a job as an architectural worker. (　)

3. *Watch the video for the third time and fill in the following blanks.*

As a construction _____, your job is to find a job to _____, find the substrates that you need to combine all the _____, analyze all the _____, read the specifications, make sure you're completely covered and show the owner a summary of all the _____ for the construction, and show them that here's the _____ cost for the construction. A construction estimator is not just about reading _____, but putting the project in their mind and _____ it in their mind and the knowledge of how a building or _____ goes together. It's not just about cost at that point; it is about how costs and changes to the _____ influences cost.

4. *Share your opinions with your partners on the following topic for discussion.*

1) How do you feel the day of being a construction estimator? Please summarize the features.

2) Can you use a few lines to list what's your understanding about building cost? Please use an example to clarify your thoughts.

Task Two Cost estimating

New Words

estimating n. 评估

subcontractor n. 转包商

survey n. 测量

excavation n. 挖掘

loam n. 壤土,肥土

landscaping n. 对……做景观美化

footing n. 基础

foundation n. 地基

patio n. 院子,天井

miscellaneous adj. 混杂的

built-in n. 嵌入式

视频链接及文本

Exercises

1. *Watch the video for the first time and choose the best answers to the following questions.*

  1) The purpose of this video is to _____.

    A. provide a cost estimating tool that can help you accurately figure the costs of your next home building project

    B. introduce the website for your building advisory

    C. make sure you are not forgetting any figure

    D. create more free content for your home building needs

  2) The five major sections includes the following EXCEPT _____.

    A. pre-construction   B. site work

    C. subcontractors   D. post-construction

  3) The third section includes _____.

    A. concrete work   B. site survey

    C. landscaping   D. dry walls

  4) The miscellaneous includes _____.

    A. tree removal   B. contingency amount

    C. landscaping   D. roofing

  5) This passage aims to _____

    A. help people to estimate home building project

    B. introduce a free estimate tool

    C. help people get exactly what they want at the price they want

    D. advertise an estimator

2. *Watch the video again and decide whether the following statements are true or false.*

  1) The video in question is about the cost of home construction projects.(   )

  2) This video will give you a cost-estimation tool for paying.(   )

  3) The PDF Cost Estimation Tool also allows you to place three different estimates side by side, so that when you collect each new estimate, you can't visually see the difference.(   )

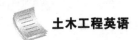

4) The last part is miscellaneous, including detailed lists, dry walls, paint, roofs and frames.(   )

5) All you have to do is click on the link below to download your free copy in the description section of this video.(   )

3. *Watch the video for the third time and fill in the following blanks of the table.*

This tool also allows you to put three different _____ side by side, so you can't visually see the _____ as you collect each new estimate. Let's take a look at that cost estimating. So I've broken down the cost estimator into five major _____. The first being _____, including things like site, _____, design, and structural engineering, and so on. The next section is site work, including things like _____, loam and seed, tree removal…things like _____. The next section then would be _____ stone brick work. Things in here would be _____ and foundations, patios and fireplaces. Next is subcontractors which include the extensive _____ of possible subcontractors, dry walls, painting, roofing, and framing.

4. *Share your opinions with your partners on the following topic for discussion.*

1) Do you know the ways of building cost control ?

2) Can you allocate the cost of construction effectively?

IV. Talking

Task One   Classical Sentences

Directions: *In this section, some popular sentences are supplied for you to read and to memorize. Then, you are required to simulate and produce your own sentences with reference to the structure.*

General Sentences

1. If it doesn't rain tomorrow, I think I'll go shopping.
   如果明天不下雨,我想去购物。

2. There's a possibility we'll go, but it all depends on the weather.
   我们有可能去,但要看天气怎么样。

3. Let's make a date to go shopping next Thursday.
   我们约好下周四去购物吧。

4. If I have time tomorrow, I think I'll get a haircut.
   如果我明天有空,我想去剪头发。

5. I hope I remember to ask the barber not to cut my hair too short.
   我希望我记得叫理发师不要把我的头发剪得太短。

6. If I get my work finished in time, I'll leave for New York on Monday.
   如果我能及时完成工作,我周一就去纽约。

7. Suppose you couldn't go on the trip, how would you feel?
   设想一下，如果你不能去旅游，你会有什么感觉？

8. What would you say if I told you I couldn't go with you?
   如果我告诉你我不能和你一起去，你会怎么想？

9. If I want to buy the car, I'll have to borrow some money.
   如果我想买那辆车的话，我就得借些钱。

10. We may be able to help you in some way.
    我们也许能在某些方面帮助你。

11. If you were to attend the banquet, what would you wear?
    如果你要参加宴会，你会穿什么？

12. What would you have done last night if you hadn't had to study?
    如果昨天晚上你不用学习的话，你会做什么？

13. I would have gone on the picnic if it hadn't rained.
    要不是之前下雨了，我就去野炊了。

14. If you had gotten up earlier, you would have had time for breakfast.
    如果你早一点起床，就有时间吃早饭了。

15. If I had had time, I would have called you.
    我要是有时间，就给你打电话了。

16. Would he have seen you if you hadn't waved to him?
    要是你没向他挥手，他还能看见你吗？

17. If he had had enough money, he would have bought that house.
    他要是有足够的钱，就会买下那房子了。

18. I wish you had called me back the next day, as I had asked you to.
    可惜你没有按我的要求，在第二天给我回个电话。

19. If you hadn't slipped and fallen, you wouldn't have broken your leg.
    如果你没滑跌倒，你就不会摔断了腿。

20. If I have known you want to go, I would have called you.
    要是我知道你想去，我就叫你了。

21. Had I known you didn't have the key, I wouldn't have locked the door.
    要是我知道你没有钥匙，我就不会锁门了。

22. She would have gone with me, but she didn't have time.
    她本想和我一起去的，可是她没时间。

23. If I had asked directions, I wouldn't have got lost.
    要是我问一下路，就不会走丢了。

24. Even if we could have taken the vocation, we mightn't have wanted to.
    即使我们可以休假，我们也许不想去呢。

25. Everything would be alright, if you had said that.
    如果你说过，一切都好办了。

26. Looking back on it, I wish we hadn't been given in so easily.
现在回想起来,我真希望我们没有那么轻易地让步。

27. One of these days, I'd like to take a vacation.
总有一天,我要去休假。

28. As soon as I can, I'm going to change jobs.
我要尽快换个工作。

29. There's a chance he won't be able to be home for Christmas.
他可能不能回家过圣诞节了。

30. What is it you don't like the winter weather?
你为什么不喜欢冬天的天气?

31. I don't like it when the weather gets really cold.
我不喜欢天太冷。

32. The thing I don't like about driving is all the traffic on the road.
我不喜欢开车去是因为路上很拥挤。

33. He doesn't like the idea of going to bed early.
他不喜欢早睡。

34. I like to play tennis, but I'm not a very good player.
我很喜欢打网球,但是打得不是很好。

35. I don't like spinach even though I know it's good for me.
我不喜欢菠菜,尽管我知道菠菜对我有好处。

36. I'm afraid you're being too particular about your food.
恐怕你对食物太挑剔了。

37. He always finds fault with everything.
无论是什么事情,他总会找碴。

38. You have wonderful taste in clothes.
你对衣服很有品味。

39. What's your favorite pastime?
你最喜欢的消遣是什么?

40. What did you like best about the movie?
你最欣赏这部电影的哪个方面?

41. The feature started at 9 o'clock and ended at 11:30.
专题片从九点开始,一直持续到十一点半。

42. They say the new film is an adventure story.
他们说这部新影片讲的是一个冒险故事。

43. We went to a concert last night to hear the symphony orchestra.
我们昨天晚上去音乐厅听交响乐了。

44. A group of us went out to the theater last night.
昨晚我们一群人去了剧院。

45. The new play was good and everybody enjoyed it.
   这个新剧很好, 人人都喜欢。

46. By the time we got there, the play had already begun.
   我们到达时戏已经开始了。

47. The usher showed us to our seats.
   引导员把我们带到了座位前。

48. The cast of the play included a famous actor.
   这场戏的演员阵容里有一位很著名的演员。

49. After the play was over, we all wanted to get something to eat.
   戏结束了以后, 我们都想去吃点东西。

Specialized Sentences

1. A cost estimate at a given stage of project development represents a prediction provided by the cost engineer or estimator on the basis of available data.
   在项目开发过程中, 某一特定阶段的成本估算就是造价工程师或造价员根据现有的数据进行的预测。

2. These basic costs may then be allocated proportionally to various tasks which are subdivisions of a project.
   这些基本成本将会按比例分配到工程子项目的不同任务中。

3. Construction cost only constitutes a fraction, though a substantial fraction, of the total project cost.
   建筑成本也只是工程成本的一部分, 尽管是很大一部分。

4. A unit cost is assigned to each and then the total cost is determined by summing the costs incurred in each task.
   一旦这些工作及其工作量确定下来, 每个工作将能确定一个单价, 然后将每个工作发生的费用相加即可确定总费用。

5. The allocation of joint costs should be related to the category of basic costs in an allocation process.
   综合成本的分配应该与基本成本的分配相关。

6. Erection of the equipment will be carried out according to the specifications and drawings.
   设备安装将按照说明书和图纸进行。

7. All site erection works will be performed by the buyer under the technical instruction of the seller.
   所有的现场建造工作都应在卖方的技术指导下由买方完成。

8. The construction company is fully in charge of the administration of all erection work.
   建设公司完全负责全部建造工程的管理。

9. Our company cover all construction activities, that is: piling, civil engineering, mechanical erection, piping, electrical, instrumentation, painting and insulation work.

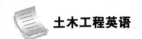

我们公司涉及所有施工活动,包括:打桩、土建工程、机械安装、配管、电气、仪表、油漆和保温工程。

10. What is the feature of this cracker (cracking furnace, heating furnace, reactor, mixer, centrifuge and belt-conveyer)?

这台裂解器(裂解炉、加热炉、反应器、搅拌器、离心机以及皮带输送机)的特点是什么?

11. The spherical tank (gas holder, container) will be shipped in the condition of edge prepared and bent plates.

这个球形箱(储气罐、容器)将以板加工和弯板的形式装运。

12. I think that the on-site training will be necessary for the tank.

我认为对罐装工作进行现场培训是十分必要的。

13. The cooler (condenser, separator, boiler, generator, scrubber, stripper, heat exchanger) is a pressure vessel. It is subject to the pressure vessel code.

这台冷却器(冷凝器、分离器、锅炉、发生器、洗涤器、汽提器、热交换器)是一个压力容器,它必须服从压力容器法规。

14. The pressure vessel must be inspected by our authoritative organization—Administration of Labor.

压力容器必须接受我们的有权机关劳动总局的监察。

15. The new reciprocating (centrifugal, opposed-balanced) compressor will be installed next week.

下周将安装这台新的往复式(离心式、对置平衡式)压缩机。

16. The distilling column (absorber, column evaporator, regenerator, column washer) is a kind of equipment for outdoor installation.

蒸馏塔(吸收塔、柱式蒸发器、蓄热器塔、洗涤塔)是一种室外安装的设备。

17. The TG70 steam turbine has an operation speed of 9,600 rotations per minute (RPM).

TG70 型蒸汽透平的运转速度为每分钟 9,600 转(RPM)。

18. What do you think of this erection work?

你看这项建造工作如何?

19. This low (middle, high) pressure blower (pump) will be assembled in the No. 3 workshop.

这台低(中、高)压鼓风机(泵)将在三号车间安装。

20. We are adjusting (installing, checking, aligning, leveling, purging) the equipment.

我们正在调整(安装、检查、校准、找平、清洗)这台设备。

21. The working team will finish the job next week.

工作班组将在下周干完这活。

22. We can adjust the levelness of the machine by means of shim and screw jack.

我们可以利用垫片和千斤顶来调整机器的水平度。

23. After seven days, the grouted mortar will have concreted, then we shall tighten the

anchor bolts.

七天以后灌浆凝固,我们就将拧紧地脚螺栓。

24. The alignment of the coupling should be performed by two dial gauges.

通过两只千分表可以校准联轴器。

25. The maximum allowable misalignment of the coupling is 0.02mm.

联轴器的最大允许偏差为 0.02 毫米。

26. How many radial (axial) clearance are there in this bush (journal bearing, thrust bearing)?

这个轴套(轴颈轴承、止推轴承)的径向(轴向)间隙是多少?

27. Does the bolt fit the nut?

螺栓与螺母匹配吗。

28. We prefer welding to riveting.

我们认为焊接比铆接好。

29. Do you know how to assemble (adjust) this new machine?

你知道如何装配(调整)这台新机器吗?

30. Total Quality Control(TQC) is a better quality control system.

全面质量管理(简称 TQC)是一种较好的质量管理体系。

31. TQC over the project will be strengthened.

对于这个工程的全面质量管理将要加强。

32. To maintain the best quality of the construction work is the important responsibility of the field controllers.

确保施工质量优良是现场管理人员的重要职责。

33. We possess skilled technician and complete measuring and test instruments used to ensure the quality of engineering.

我们拥有熟练的技术力量和齐全的检测手段,可以确保工程质量。

34. Field inspection work is handled (executed, directed) by our Inspection Section.

现场检查工作由我们的检查科管理(实施、指导)。

35. Our site quality inspector will report to the Project Manager every day.

我们的现场质量检查员将每天向工程项目经理汇报。

36. I want to see the certificate of quality (certificate of manufacturer, certificate of inspection, certificate of shipment, material certificate, and certificate of proof).

我要看看质量证书(制造证书、检查证明书、出口许可证书、材料合格证、检验证书)。

37. Here is the report of chemical composition inspection.

这是化学成份检验报告。

38. Is it OK (good, guaranteed, satisfied, passed)?

那是正确的(好的、保证的、满意的、合格的)吗?

39. We shall take the sample to test its physical properties (mechanical properties, tensile

strength, yield point, percentage elongation, reduction of area).

我们将取样试验其物理性能(机械性能、抗拉强度、屈服点、延伸率、断面收缩率)。

40. We have received certificate of Authorization for the fabrication and erection of pressure vessels.

我们具有压力容器制作和安装的授权认可证书。

41. The welds passed the examination of radio graphic test (ultrasonic inspection, magnetic testing).

这焊缝通过射线透视检查(超声波探伤、磁力探伤)是合格的。

42. Are you a qualified nondestructive testing (NDT) person?

你是具有资格的无损检测人员(NDT)吗?

43. Let us go to the laboratory to check the radio graphic films.

请到试验室去检查透视片子。

44. This job will have to be done over again.

这项工作必须返工重做。

45. The defect must be repaired at once.

缺陷必须立即修理。

46. This problem of quality needs a further discussion.

这个质量问题需要进一步研讨。

47. The ISO standards have been used by our company in this project.

我公司已在本工程中使用国际标准(ISO)。

48. The testing results fulfill quality requirement.

试验结果达到质量要求。

49. Check list (quality specification) has been signed by the controller (inspector, checker).

检验单(质量说明书)已由管理员(检查员、审核人)签字。

50. The standards "GB" and "YB" provide the method of testing for materials in our country just like the standard ASTM in America.

我国的国标 GB 和冶金标准 YB 中规定了材料的试验方法,正如美国的 ASTM 标准一样。

Task Two　Sample Dialogue

Directions: *In this section, you are going to read several times the following Sample Dialogue about the relevant topic. Please pay special attention to five C's (culture, context, coherence, cohesion and critique) in the Sample Dialogue and get ready for a smooth communication in the coming task.*

Building Camps

(*To construct the works smoothly and efficiently, the site needs to be arranged properly. Today, Mr. Bian and Mr. Cheng are talking to Mr. Sukh about the plan to lay out the whole site.*)

Sukh: The other day, you told me that you were going to build up two camps on the site. Could you elaborate on your plan, Mr. Bian?

Bian: According to our construction plan, our first camp, which we call "Camp A", will be located at the junction of the two access roads leading to the dam and the power house. It will mainly be used for storage of construction materials, spare parts, etc... A parking lot and a repair shop will be set up within this area. The power bender will also be located here.

Sukh: What will be the size of this camp?

Bian: It will cover an area of about 4,000 square meters.

Sukh: How much is that in acres? I am not quite familiar with the Metric System.

Bian: Sorry, we are not used to the British System.

Cheng: Just a minute. Let me work it out with my small calculator. Let's see,...It's 4,785 square yards. About one acre.

Sukh: That's acceptable. This piece of land belongs to the government. We'll get it approved soon. What about the other camp?

Bian: The other camp, which we call "Camp B", is mainly for the accommodation of the Chinese professionals and the offices. It's also where we have our weekly and monthly review meetings. What do you call this kind of camp?

Cheng: Man camp, to be specific, but it belongs to construction camp in general; or you could say it's part of the whole construction camp.

Sukh: There's no clear distinction between these terms, even among professionals. When will it be located?

Bian: On the right-hand side of the access road, about half a mile from Camp A. It will occupy approximately the same area.

Sukh: Why do you have to choose this area, Mr. Bian? It is a part of this Country's nature preserve. Though our government is in full support of this project, it won't sanction it at the price of damaging its beautiful environment. I don't think we could get the approval from the appropriate department.

Bian: You see, this area is ideal for Camp B. It's comparatively flat and this makes it easy for us to build the houses for bedrooms and offices. The key point is, it's close to the job site and the source of a portable water supply. Once we are given that piece of land, we'll be strictly confined to it and do everything we can to keep it in good order.

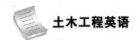

Anyway, could you have a try with the government, please?

Sukh: OK, if you insist. But I don't see much chance of its being approved.

Bian: Now, I would like Mr. Cheng to describe our plan to arrange the other temporary facilities on the site.

Cheng: From our site investigations, we have found a piece of land with a gentle slope, where we want to locate the batching plant and the crushing system because it is very close to the dam site and easy to be flattened.

Sukh: What's the exact distance from the dam site and in which direction?

Cheng: It's 500 meters, oh, around 1,630 feet in your system, to the south of the dam site, on the right bank of the River. If we can get the approval, it will facilitate our concrete production and transportation.

Sukh: Will you submit to us all the layouts of the temporary works together with a letter of application? We'll try to get them approved soon, if possible.

Bian: The sooner, the better.

Sukh: I will do my best, but I can't commit myself to getting the approval. All I can promise is that you will have a reply within a week, whether positive or negative.

Bian: That's what the contract says!

Task Three   Simulation and Reproduction

Directions: *The class will be divided into three major groups, each of which will be assigned a topic. In each group, some students may be the teacher, while others may be students. In the process of discussion, please observe the principles of cooperation, politeness and choice of words. One of the groups will be chosen to demonstrate the discussion to the class.*

1) Layouts of the temporary works of certain project in our school or our city.

2) A story related to building camps on the site.

3) The importance of professional cost estimating.

Task Four   Discussion and Debate

Directions: *The class will be divided into two groups. Please choose your stand in regard to the following controversy and support your opinions with scientific evidences. Please refer to the specialized terms and classical sentences in the previous parts of this unit.*

To make a good estimate, estimators would look at other bids they've done for other projects and incorporate those into ours; bids for equipment, suppliers, rooftop units, pumps, air inline units, exhaust fans, supply grills-things like that. Do you agree?

V. After-class Exercises

1. *Match the English words in Column A with the Chinese meanings in Column B.*

A	B
1) estimator	A) 成本估算
2) unit cost	B) 预算控制
3) stress	C) 分配
4) cost estimation	D) 载荷
5) budget control	E) 单价
6) production function	F) 造价师
7) load	G) 刚度
8) allocation	H) 产出函数
9) rigidity	I) 抑制
10) restrain	J) 应力

2. *Fill in the following blanks with the words or phrases in the word bank. Change the forms if it's necessary.*

unit	transformation	property	damper	concrete
cost engineer	controllers	erection	requirement	allocation

1) A cost estimate at a given stage of project development represents a prediction provided by the _____.

2) A unit cost is assigned to each and then the _____ is determined by summing the costs incurred in each task.

3) Manufacturing can be defined as the _____ of raw materials into useful products through the use of the easiest and least-expensive methods.

4) The construction company is fully in charge of the administration of all _____ work.

5) We shall take the sample to test its physical _____.

6) The testing results fulfill quality _____.

7) Though frame material and design should handle damping, _____ are sometimes built into frame sections to handle specific problems.

8) Most frames are made up of cast iron, weld steel, composition, or _____.

9) The _____ of joint costs should be causally related to the category of basic costs.

10) To maintain the best quality of the construction work is the important responsibility of the field _____.

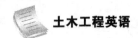

土木工程英语

3. *Translate the following sentences into English.*
   1）某一特定阶段的成本估算就是造价工程师根据现有的数据进行的预测。

   _____

   _____

   2）这些基本成本将会被按比例分配到工程子项目的不同单元中。

   _____

   _____

   3）建筑成本只是构成工程成本的一部分,尽管是很大一部分。

   _____

   _____

   4）每个工作将能确定一个单价,然后将每个工作发生的费用相加即可确定总费用。

   _____

   _____

   5）综合成本的分配应该与基本成本的分配有关。

   _____

   _____

4. *Please write an essay of about* 120 *words on the topic：Application of Cost Estimating in Building. Some specific examples will be highly appreciated and watch out the spelling of some specialized terms you have learnt in this unit.*

   _____

   _____

   _____

   _____

   _____

   _____

   _____

VI. Additional Reading

Construction Cost Estimate

An accurate construction cost estimate is crucial to a successful construction project. An accurate construction cost estimate can tell how long a project will take and how much it would cost. However, getting an accurate construction cost estimate can be hard. And the consequences of an overestimate or underestimate can detrimentally harm projects. The amount of time and effort put into a construction cost estimate saves businesses money and time in the end. While the systems are more complex, the methodology for performing a construction cost estimate is fairly easy. This brief guide can help you perform a construction cost estimate.

Construction Cost Estimation Overview

The American Society of Professional Estimators lays out five levels of system estimates. These levels of estimates also correspond(符合) to a level of accuracy. As more data comes in at the other levels, the estimates become more accurate, while ensuring it is the right program for them. After the five levels of estimates, there are the three types of estimates, they correspond to the various levels and project stage.

Design estimates are a construction cost estimate on the design stage of a project. They use the Order of Magnitude level, Schematic Design level and Design Development level. The Order of Magnitude determines feasibility before the project design starts. Schematic Design estimate uses the schematic design to estimate costs and help determine feasibility. And the Design Development and Construction Document phases use the engineer's estimate and construction documents.

The Bid Estimate phase presents the bid. It uses multiple data points such as construction documents and other direct costs. With these data points, the estimator determines an approximation of what the job should cost and submits it with the other paperwork ensuring them the potential to work on the project.

Creating Construction Cost Estimates

Creating a construction cost estimate might seem daunting(令人畏惧的). And when considering that cost estimating used to all be done by hand, it can seem impossible. However, nowadays there are systems and tools that make a construction cost estimate easier. The Uniform at System for building estimation is a government standard for estimating buildings. It starts with the major group elements then drills deeper into individual elements. This helps break up building estimates so they're in easy to understand and complete parts.

Besides dividing up cost estimation into specific groups, estimators must include a bill of quantities. The bill of quantities is an itemized list of work and materials necessary for the project. For a construction cost estimate, that would include take-off quantities and more to get the accurate number. Estimators use elements like construction documents, squaring and abstracting in order to come up with the appropriate numbers for billing.

Construction Cost Estimate Components

One project's construction cost estimate will use the same elements and methods as another project; however, each project is different. Usually, an estimator will look at the measure of materials and labor needed to complete the project. Sometimes they'll include a secondary look at labor hours and rates. They always look at material and equipment costs, these can include the costs of operation and repair or rent. Great estimators will also look at the indirect costs of a potential

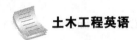

project. These can include everything from administrative costs, legal fees, permits, bonds, transportation, or storage costs. Of course, there is usually a contingency(意外开支) estimate and a look at what profit should be. These components make up a construction cost estimate.

Cost Estimation Approaches

Estimators use and gather almost all the same data in almost the same way. How they use the data in their estimates differs by their approach to construction cost estimation. Some cost estimators use unit cost estimating. When every unit of work has an associated cost it's fairly easy to put together all the data into an estimate. Another incredibly accurate measure of estimation is called stick estimating. It uses a complete list of materials, the labor schedule down to the hour, all vendor(供应商) proposals, costs, profits, all data points. Then the estimator takes the list of items and calculates the total cost and uses that as the estimate. It's incredibly accurate especially for estimators with many years in the industry.

In this information age, the tools needed for estimation are easier to find, which increases the expected accuracy of construction cost estimation. Understanding all of the elements involved in proper estimation makes it easier to get a more accurate number. Accurate estimation matters in construction because it determines whether a project will succeed on time and on budget or not. With the understanding of a basic construction cost estimate and the right tools, a more accurate estimate will be easier to come by.

1. *Read the passage quickly by using the skills of skimming and scanning. And choose the best answer to the following questions.*

    1) What can be done with accurate estimates of construction costs? Examine and verity _____.

        A. whether a project worth doing or not

        B. how many people are working on a project

        C. prospects of a project

        D. how long does a project take and how much does it cost

    2) What are the consequences of the time and effort invested in building cost estimates? _____.

        A. Waste of time

        B. Waste of money

        C. It will eventually save the company money and time

        D. Do nothing

    3) What can this short guide help users? _____.

        A. Execute the construction cost estimate   B. Avoid construction errors

        C. Save money in construction   D. Save time in construction

    4) What are the five levels of system estimates proposed by the American Institute of

Professional Estimation? _____.

   A. Useless standard   B. Very simple codes

   C. Certain levels of accuracy   D. Inaccurate estimates

5) What's the design estimate? _____.

   A. To estimate the amount of money spent on construction

   B. To estimate construction costs for the project design phase

   C. Estimation of time

   D. Estimation of market prospects

6) What techniques are used to estimate construction costs? _____.

   A. Order of magnitude   B. Principle Design Level

   C. Design development level   D. The above three options are correct

7) Is it possible to create a construction cost when considering that all cost estimates are done manually? _____.

   A. Impossible   B. unrealistic

   C. Unable to complete   D. Possible

8) In addition to divide the cost estimate into specific groups, what must be included in the estimate? _____.

   A. Workers' wages   B. Bill of quantities

   C. Exchange rate   D. Tax

9) What does the Estimator use to arrive at the appropriate billing figures? _____.

   A. Access to information

   B. Corporate documents

   C. Takeoff, construction documentation, segmentation and abstraction

   D. Everything they can use

10) What increases the accuracy of projected construction cost estimates? _____.

   A. Variety of projects

   B. An increase in profits

   C. In this information age, it is easier to use the tools needed

   D. Understanding the basic construction cost estimates

2. *In this part, the students are required to make an oral presentation on either of the following topics.*

1) The ways of cost estimating.

2) The function of cost control.

习题答案

Unit Seven Big Ben

I. Pre-class Activity

Directions: *Please read the general introduction about **Big Ben** and tell something more about the great architecture to your classmates.*

Big Ben

Big Ben is the nickname for the Great Bell of the clock at the north end of the Palace of  Westminster in London and is usually extended to refer to both the clock and the clock tower. The official name of the tower in which Big Ben is located was originally the Clock Tower, but it was renamed Elizabeth Tower in 2012 to mark the Diamond Jubilee of Elizabeth II.

The tower was designed by Augustus Pugin in a neo-gothic style. When completed in 1859, its clock was the largest and most accurate four-faced striking and chiming clock in the world. The tower stands 315 feet (96 m) tall, and the climb from ground level to the belfry is 334 steps. Its base is square, measuring 39 feet (12 m) on each side. Dials of the clock are 23 feet (7.0 m) in diameter. On 31 May 2009, celebrations were held to mark the tower's 150th anniversary.

Big Ben is the largest of five bells and weighs 13.5 long tons (13.7 tons; 15.1 short tons). It was the largest bell in the United Kingdom for 23 years. The origin of the bell's nickname is open to question; it may be named after Sir Benjamin Hall, who oversaw its installation, or heavyweight boxing champion Benjamin Caunt. Four quarter bells chime at 15, 30 and 45 minutes past the hour and just before Big Ben tolls on the hour. The clock uses its original Victorian mechanism, but an electric motor can be used as a backup.

The tower is a British cultural icon recognized all over the world. It is one of the most prominent symbols of the United Kingdom and parliamentary democracy, and it is often used in the establishing shot of films set in London. The clock tower has been part of a Grade I listed building since 1970 and a UNESCO World Heritage Site since 1987.

On 21 August 2017, a four-year schedule of renovation works began on the tower, which are include the addition of a lift. There are also plans to re-glaze and repaint the clock dials. With a few exceptions, such as New Year's Eve and Remembrance Sunday, the bells are to be silent until the work has been completed in the 2020s.

II. Specialized Terms

Directions: *Please memorize the following specialized terms before the class so that you will be able to better cope with the coming tasks.*

access n. 交通, 访问, 入门

agreement n. 同意, 协议

allocation n. 分配

approval n. 同意, 批准

arbitration n. 仲裁

Asia Development Bank 亚洲开发银行

assistant n. 助理, 助手

authorize (delegate) v.授权

bill of quantities (boq)工程量表

civil works 土建工程

claim v.索赔

comment n. 评论, 意见

commercial manager 商务经理

conditions of contract 合同条件

general conditions 通用条件

conditions of particular application 专用条件

special conditions of contract 合同特别条款

conditions of contract for works of civil engineering construction 土木工程施工合同条款

construction management 施工管理

consultant n. 顾问

contract agreement 合同协议

contractor n. 承包商

cooperation n. 合作

coordination n. 协调

cost n. 费用

cost control 成本控制

counterclaim v.反索赔

delay v.延误

demobilization n. 退场

department n. 部门

designer n. 设计者

drawing n. 图纸

shop drawing 施工图

design drawing 设计图

as-built drawing 竣工图

blue drawing 蓝图

transparent drawing 透式图

construction drawing 施工图

electric works 电气工程

employer (client, owner) n. 业主

engineer n. 工程师

engineer's representative 工程师代表

engineering project 工程项目

international project 国际工程

overseas project 海外工程

domestic project 国内工程

equipment n. 设备

expatriate n. 外籍职员

expert n. 专家

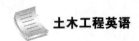

export n. 出口

federation of civil engineering contractor 土木工程承包商联合会

formal n. 正式的

hydromechanical works 水力工程

in charge of 负责，主管

informal adj. 非正式的

institution of civil engineers 土木工程协会

instrument n. 仪器，器械

insurance n. 保险

labour n. 劳务

layout v.布置

leading company (sponsor)牵头公司

liability (responsibility, obligation) 责任

lump sum 一次性付款

machinery n. 机械

manpower n. 人力资源

manufacturer 制造商

material n. 材料

measure n. 办法，措施

take effective measures 采取有效措施

measurement n. 测量，计量

memorandum n. 备忘录

mobilization n. 进场

objection n. 反对

payment n. 支付

plant n. 设备

point of view (opinion)观点，意见

prequalification n. 资格,预审

procurement n. 采购

profit n. 利润

progress control 进度控制

project manager 项目经理

quality control 质量控制

request (application) v.审查

risk n. 风险

river closure 截流

river diversion 导流

safety n. 安全

signature n. 签名

site n. 工地,现场

site engineer 现场工程师

specification n. 规范

staff n. 职员

subcontractor n. 分包商

submission n. 提交

supervise vt. 监督,监视

supplier n. 供贷商

the international chamber of commerce 国际商会

variation n. 变更

III. Watching and Listening

Task One　Big Ben（Ⅰ）

New Words

iconic adj. 符号的；偶像的

mega adj. 宏大的；精彩的

blueprint n. 蓝图；设计图

skyscraper n. 摩天大楼

blow apart 炸开

dismantle vt. 拆卸,拆开

cog n. 轮齿

vault n. 拱顶,穹窿

labyrinth n. 迷宫；错综复杂

medieval castle 中世纪城堡

视频链接及文本

spectacular adj. 壮观的	intricate adj. 错综复杂的
spire n. 尖塔	dial n. 表盘
marvel n. 奇迹	wristwatch n. 手表
restoration v.复原,恢复	gust n. 狂风
ornate adj. 装饰华丽的	sail n. 帆船
pioneer n. 先驱者	swing v.摇摆
landmark n. 地标	pendulum n. 摇摆不停
mechanism n. 装置; 结构	ratchet n. 棘齿

Exercises

1. *Watch the video for the first time and choose the best answers to the following questions.*

   1) Standing on the banks of River _____, in the heart of London, England, is the palace of Westminster.

     A. Tops   B. Thomas

     C. Tomes   D. Thames

   2) This immense palace was completed in _____.

     A. 1780   B. 1870

     C. 1770   D. 1880

   3) This immense palace has three spectacular towers to the _____.

     A. north   B. west

     C. south   D. east

   4) The _____ Tower, the tallest square tower in the world when it was completed.

     A. Victoria   B. Big Ben

     C. Westminster   D. parliament

   5) You've got _____ stairs in the Clock tower and no left, unfortunately.

     A. 434   B. 403

     C. 304   D. 334

2. *Watch the video again and decide whether the following statements are true or false.*

   1) The only way to solve these mysteries is to blow this immense palace apart. (　)

   2) Big Ben is better known as the Houses of parliament. (　)

   3) A labyrinth of more than 30 mile of corridors connecting more than 1,000 grand rooms, chambers and vaults. (　)

   4) It has more than 1,000 rooms spread across 18 acres. (　)

   5) Inside this iconic clock tower hangs Big Ben itself, the nearly 14 tons bell that gives the tower its nickname. (　)

3. *Watch the video for the third time and fill in the following blanks of the table.*

The bells that _____ on the quarter-hour sit around it. Their _____ are controlled by the intricate ticking mechanism at the heart of the tower. It keeps the clock _____ to within just 2

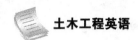

seconds a week. On each side, 312 shard of opal glass make up the clock _____. The copper _____ hand, are l feet long as tall as a _____ docker bus. These famous _____ have kept Landon running on time for 1,600 years. Like any ordinary wristwatch, this clock needs _____. Just as it did in Victorian times. Put the winding handle on, and then winding for about the next hour and a _____. When parliament commissioned this clock in 1854, they _____ Big Ben be the biggest most powerful Chiming clock in the world.

4. *Share your opinions with your partners on the following topic for discussion.*

　　1) Do you know Big Ben? How do you know this famous structure?

　　2) Can you use a paragraph to illustrate what kind of structure keeps the world's most famous clock in perfect time?

Task Two　Big Ben (Ⅱ)

New Words

视频链接及文本

tick v.发出滴答声

triumph n. 胜利；巨大成就

magnificent adj. 壮丽的

tile v.用瓦片……覆盖

destabilize v.失去稳定性

colossal adj. 巨大的

subterranean adj. 地下的

tunnel n. 隧道；地道

scenario n. 情景

collapse v.倒塌

excavate vt. 挖掘，发掘

detect v.查明；发现

gravel n. 砾石,砂砾

lean v.倾斜；倚靠

shaft n. 通道

valve n. 阀

crack n. 裂缝

precision n. 精确度

inject v.注入

ingredient n. 因素；材料

ingenious adj. 精巧的

upright adj. 竖直的

pump n. 泵

Exercises

1. *Watch the video for the first time and choose the best answers to the following questions.*

　　1) Made for over 1 million stones and bricks, this tower is enormously heavy. _____ tons of iron tile sit on top of the spire.

　　　　A. 18　　　　　　　　　　　B. 28

　　　　C. 38　　　　　　　　　　　D. 48

　　2) The Victorians built a _____ foundation almost 10 feet thick to stop the towers from sinking into the wet sand beneath.

　　　　A.wooden　　　　　　　　　B. steel

　　　　C. concrete　　　　　　　　　D. stone

　　3) Today the construction of the new _____ poses a constant threat to buildings on the surface.

A. railways   B. train tunnels

C. train stations   D. ways

4) Geoff Morris knew the worst-case scenario would be _____.

A. collapse   B. leak

C. detect   D. the parliament

5) As tunneling progressed, engineers were alarmed to _____ movement in the buildings above.

A. find   B. look

C. think about   D. detect

2. *Watch the video again and decide whether the following statements are true or false.*

1) The last thing you want is structures above ground moving because of what you're doing below ground. (   )

2) There was a problem with the ground movement because of the foundations becoming stabilized.(   )

3) If you've been to the seaside and you try to dig a hole, it just instantly fills up with water and grows deeper instead of wider.   (   )

4) The earth under Big Ben is extremely stable. (   )

5) For over a century, Big Ben has been the ticking heart at the City's center. (   )

3. *Watch the video for the third time and fill in the following blanks.*

As the 24-foot-wide train tunnels ran close to Big Ben's foundations, the tower begin to _____. To keep it stable, engineer dug a deep _____, and push long steel tubes into the ground right beneath Big Ben. Each tube had 600 valves, so workers could _____ in a concrete mixture to fill the _____ in the ground with precision. As the engineers dug the tunnels, they _____ over 220 tons of concrete mixture, successfully _____ Big Ben from becoming a leaning _____ of Pisa. Once they _____ the ground, it would've stopped any potential further movement on the clock tower. One of the _____ ingredients into successful tunneling. This ingenious intervention kept the tower standing _____ and the bells ringing on time.

4. *Share your opinions with your partners on the following topic for discussion.*

In 1994, the London underground excavated new train tunnels just 90 feet from the base of Big Ben. As tunneling progressed, engineers were alarmed to detect movement in the buildings above. What should they do to keep this structure stable?

IV. Talking

Task One   Classical Sentences

Directions: *In this section, some popular sentences are supplied for you to read and to memorize.*

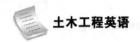

Then, you are required to simulate and produce your own sentences with reference to the structure.

General Sentences

1. Would you please tell Mr. John that I'm here?
   请问你能告诉约翰先生我在这吗?

2. Would you help me lift this heavy box?
   你能帮我将这个重盒子抬起来吗?

3. Please ask John to turn on the lights.
   请让约翰把灯打开。

4. Get me a hammer from the kitchen, will you?
   从厨房里给我拿个锤子,好吗?

5. Would you mind mailing this letter for me?
   你愿意帮我发这封邮件吗?

6. If you have time, will you call me tomorrow?
   如果你有时间,明天给我打电话好吗?

7. Please pick up those cups and saucers.
   请将那些杯子和碟子收拾好。

8. Will you do me a favor?
   你能帮个忙吗?

9. Excuse me, sir. Can you give me some information?
   先生,打扰一下。你能告诉我一些信息吗?

10. Do you happen to know Mr. Cooper's telephone number?
    你知道库珀先生的电话号码吗?

11. Would you mind giving me a push? My car has stalled.
    你能帮我推推车吗? 我的车抛锚了。

12. Would you be so kind as to open this window for me? It's stuffy.
    你能帮我把窗户打开吗? 好闷人。

13. If there's anything else I can do, please let me know.
    如果还有我能做的事情,请告诉我。

14. This is the last time I'll ever ask you to do anything for me.
    这是最后一次我麻烦你为我办事了。

15. I certainly didn't intend to cause you so much inconvenience.
    我真不想给你带来这么多不便。

16. Would you please hold the door open for me?
    请帮我开门好吗?

17. You're very kind to take the trouble to help me.
    你真是太好了,不嫌麻烦来帮我。

18. I wish I could repay you somehow for your kindness.
但愿我能以某种方式报答你的好意。

19. I'm afraid it was a bother for you to do this.
做这件事恐怕会给你带来很多麻烦。

20. He'll always be indebted to you for what you've done.
对你所做的事情,他总是感激不尽。

21. Could you lend me ten dollars? I left my wallet at home.
能借我十美元吗？我把钱包忘在家里了。

22. I'd appreciate it if you would turn out the lights. I'm sleepy.
如果你能将灯关掉的话,我会很感激。我好困。

23. You're wanted on the telephone.
有你的电话。

24. What number should I dial to get the operator?
我想接通接线员的电话,我应该拨打什么号码？

25. The telephone is ringing, would you answer it, please?
电话铃响了,你能接下电话吗？

26. Would you like to leave a message?
你想留下什么口信儿吗？

27. I have to hang up now.
我现在得挂电话了。

28. Put the receiver closer to your mouth. I can't hear you.
将话筒靠近你的嘴巴,我听不见你的声音。

29. Would you mind calling back sometime tomorrow?
你介意明天再打过来吗？

30. I almost forgot to have the phone disconnected.
我差点忘记挂断电话了。

31. It wasn't any bother. I was glad to do it.
一点儿都不麻烦,我很乐意做这件事。

32. There's just one last favor I need to ask of you.
还有最后一件事需要你的帮助。

33. I'd be happy to help you in any way I can.
很高兴我能尽我所能地帮助你。

34. Please excuse me for a little while. I want to do something.
对不起,稍等会儿。我有点事要办。

35. I didn't realize the time had passed so quickly.
我没有意识到时间过得这样快。

36. I've got a lot of things to do before I can leave.
在我走之前,我有很多事情要做。

37. For one thing, I've got to drop by the bank to get some money.
   首先,我得去下银行取些钱。

38. It'll take almost all my savings to buy the ticket.
   买这张票,几乎花费了我所有积蓄。

39. Oh, I just remembered something! I have to apply for a passport.
   我记起一件事,我得去申请个护照。

40. It's a good thing you reminded me to take my heavy coat.
   你提醒我带件厚衣服,真是太好了。

41. I would never have thought of it if you hadn't mentioned it.
   要是你不提到它,我几乎想不起来了。

42. I'll see you off at the airport.
   我会去机场送你。

43. Let's go out to the airport. The plane landed ten minutes ago.
   我们去机场吧。飞机在十分钟前就已经着陆了。

44. There was a big crowd and we had difficulty getting a taxi.
   这里人很多,我们很难打到车。

45. They're calling your flight now. You barely have time to make it.
   他们现在在广播你的班机起飞时间。你勉强来得及赶上。

46. You'd better run or you're going to be left behind.
   你最好跑过去,不然就会被落下了。

47. Don't forget to call us to let us know you arrived safely.
   你安全到达后,别忘记打电话报个平安。

48. I'm sure I've forgotten something, but it's too late now.
   我确定我忘记了什么东西,但是现在太晚了。

49. Do you have anything to declare for customs?
   你有什么需要报关的吗?

50. You don't have to pay any duty on personal belongings.
   你的私人物品不用交税。

Specialized Sentences

1. The sizes of footing are determined by dividing the loads to be imposed at the base of the footing by the allowable bearing pressure which can be imposed on the soil or rock of the earth.
   基础的尺寸,是由可能施加在基础底部的荷载除以地基和岩石能够承当的容许支撑力来确定的。

2. Retaining walls are those walls subject to horizontal earth pressures due to the retention of earth behind them.
   挡土墙是指由于泥土滞留而承受水平压力的墙。

3. Mat or raft foundations are large, thick, and usually heavily reinforced concrete mats which transfer loads from a number of columns or columns and walls to the underlying soil or rock.

板式基础或筏式基础通常是指把柱子或者柱子和墙上的荷载传导到基础土层或岩层中的面积、厚度及配筋量都很大的钢筋混凝土板块。

4. Well points, pumping from deep wells, or pumping from sumps are methods used to dewater construction sites during foundation installation.

从深井中抽水或从集水井中抽水的井点法是在基础施工期间用于施工现场排水的方法。

5. If dewatering operations are performed in an area surrounded by existing structures, precautions must be taken to protect them, as the lowering of the groundwater may cause the soil on which they are supported to subside.

如果排水作业周围有既有建筑物包围,就必须采取预防措施来保护这些建筑物,因为降低地下水可能会引起这些建筑物的下沉。

6. Compressibility is an important soil characteristic because of the possibility of compacting the soil by rolling, tamping, vibrating, or other means, thus increasing its density and load-bearing strength.

可压缩性是泥土的一个重要特性,可以通过碾压、夯实、振动或其他方法压实泥土以增加其密实程度、提高其承载强度。

7. Shock waves also are utilized to determine the depth of bedrock by measuring the time required for the shock wave to travel to the bedrock and return to the surface as a reflected wave.

通过测定振动波传到基岩和反射波回到地面所需的时间,可确定基岩的深度。

8. Shield can be steered by varying the thrust of the jacks from left side to right side of from top to bottom, thus varying the tunnel direction left or right or up or down.

盾构可由从上到下(或从左到右)来改变千斤顶的驱动推力,这样就可以从上下、左右方向改变隧道的前进角度。

9. In large shields, an erector arm is used in the rear side of the shield to place the metal support segments along the circumference of the tunnel.

在大型盾构中,盾构后面的起重臂用来在隧道周边安装金属支撑管片。

10. A circular shield has proved to be most efficient in resisting the pressure of soft ground, so most shield-driven tunnels are circular.

实践证明,圆形盾构抵抗软土压力是最有效的,所以大多数盾构掘进的隧道都是圆形的。

11. They evaluate and work to minimize the potential settlement of buildings and other structures that stems from the pressure of their weight on the earth.

他们计算并采取措施使建筑或其他结构由于自重压力引起的沉降减少到最小。

12. They coordinate the activities of virtually everyone engaged in the work.

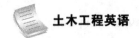

事实上,他们协调工程中每个人的工作。

13. They may also manage private engineering firms ranging in size from a few employees to hundreds.

他们也可能管理规模从几个到数百个雇员的私营建筑公司。

14. Many teaching civil engineers engage in basic research that eventually leads to technical innovations in construction materials and methods.

由于许多在基础研究领域从事教学的土木工程师的参与,会促使建筑材料和施工方法产生技术革新。

15. Piles of stone were placed at short intervals across the river, providing the bridge piers, and then a path from bank to bank was made by laying flat stone slabs across adjacent piers.

在河流中以很小的间距布置石堆作为桥墩,然后用平坦的石头横过相邻的桥墩以建成连接两岸的通道。

16. In an ordinary cantilever bridge, the gap between the ends of the cantilevers is closed, providing a continuous deck for the roadway, but if the bridge were cut in two at the point of closure each cantilever would support itself.

在普通的悬臂式桥梁中,悬臂梁端部之间的间隙是闭合的,为道路提供了连续的桥面,但是如把这种桥梁在其闭合点断开,那么每一根悬臂梁都不需额外支撑也可保持稳定。

17. Because tensile strength is not necessarily required for arch construction, arch bridges can be made of bricks or stone blocks that are held together by the compressive force characteristic of the arch.

因为拱结构不一定要求材料具有抗拉强度,所以可以用砖或石块建造拱桥,并通过拱传递压力。

18. Special attention must often be given to the design of the bridge piers, since heave loads maybe imposed on them by currents, waves, and floating ice and debris.

对桥墩的设计,要给予特殊的关注,因为桥墩要承受水流、浮冰和漂浮物而产生的撞击。

19. For constructing foundations in deep water, caissons have long been used.

在深水中建基础一般用沉箱法。

20. Workers inside the caisson, which is filled with compressed air to keep out the water, dig deeper and deeper, and the caisson sinks as the digging proceeds.

工人们在沉箱里,为隔绝水,沉箱里充满压缩空气,沉箱随着开挖的进行而下沉。

21. Stresses in a partly completed bridge—constructed by the cantilever method—can exceed the stresses in a completed bridge because of the totally different conditions of support and loading.

在用悬臂梁法施工的桥梁中,因为完全不同的支撑和荷载条件,未竣工桥梁内的应力可能会超过已竣工桥梁内的应力。

22. It also utilizes modern techniques of planning and designing new highways to allow for the habits and abilities of motorists and pedestrians and the characteristics of vehicles.

它也利用现代技术去规划、设计新型高速公路,以顺应驾车者与行人的习惯、能力及车辆的特性。

23. In the built-up parts of cities freeways generally are either depressed below ground level or elevated.

在城市建筑密集区,高速公路一般建在地下或者高架上。

24. In most cities a system of streets protect by traffic signs generally gives motorists using through streets right-of-way over cross traffic and provides positive control at intersections with other through streets.

在大多数城市中,由交通信号管理的街道系统通常给予驾车者直接通道越过横向交通的权力,为与其他直通街道相交的十字路口提供有效的管理。

25. Intersection redesign is usually aimed at eliminating bottlenecks, such as narrow pavement sections or awkward turns, or at enhancing safety by improving visibility.

道路交叉口再设计通常是为了消除道路的瓶颈问题,例如改造一些狭窄路段或急转弯,或通过改善视野提高交通安全。

26. Similarly, the use of terraced earth sheltered housing on steeply sloping hillsides can help preserve precious arable flat land in mountainous regions.

同样,在陡峭的山坡上建造梯田土保障房,有助于在多山地区保护宝贵的可耕平地。

27. This planning must consider long-term needs while providing a framework for reforming urban areas into desirable and effective environments in which to live and work.

在为把城市改造成宜居和有效的生活与工作环境时,这种规划必须考虑(城市)长期发展的需要。

28. In science fiction, future cities often are depicted as self-contained, climate-controlled units frequently located underground for protection from the elements and possibly from a hazardous or polluted environment.

在科幻小说中,常把未来城市描述成为防避风雨和躲避有害物质或污染而位于地下的设备齐全、气候可调的单元。

29. Another method of building an underwater tunnel is to sink tubular sections into a trench dug at the bottom of a river or other body of water.

另一种建造水下隧道的方法是把管段沉入江底(或其他水体底部)已经挖好的沟槽中。

30. Some long tunnels have been built with the aid of a small pilot tunnel driven parallel to the main tunnel and connected with it by crosscuts at intervals.

有些长隧道是在平行于主隧道开挖的小型导洞辅助下建造的,导洞与主隧道之间每隔一段距离由横巷连通。

31. They also complement each other in another way: they have almost the same rate of contraction and expansion.

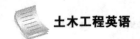

它们也以另外一种方式互补:它们几乎有相同的收缩率和膨胀率。

32. They therefore can work together in situations where both compression and tension are factors.

因此,它们在拉力、压力为主要作用因素时,能够共同工作。

33. Steel rods are embedded in concrete to make reinforced concrete in concrete beams or structures where tension will develop.

在出现拉力的混凝土梁或结构中,会将钢筋埋入混凝土从而制成钢筋混凝土。

34. Concrete and steel also form such a strong bond—the force that unites them—that the steel cannot slip with the concrete.

混凝土与钢筋形成如此强大的结合力——这个力将它们结合在一起——以至于钢筋在混凝土中不会滑移。

35. Still another advantage is that steel does not rust in concrete. Acid corrodes steel, whereas concrete has an alkaline chemical reaction, the opposite of acid.

还有另一个优势是:钢筋在混凝土中不会锈蚀。酸能腐蚀钢筋,而混凝土内部会发生碱性的化学反应,与酸相反。

36. The adoption of structural steel and reinforced concrete caused major changes in traditional construction practices.

结构钢与钢筋混凝土使传统建筑结构发生了明显的变化。

37. It was no longer necessary to use thick walls of stone or brick for multistory buildings, and it became much simpler to build fire-resistant floors.

对多层建筑,再也没必要采用厚的石墙或砖墙,且建造防火地面也变得容易得多。

38. Both these changes served to reduce the cost of construction. It also became possible to erect buildings with greater heights and longer spans.

这些变化有利于降低建筑的成本。它也使建造高度更高和跨度更大的建筑物成为可能。

39. Since the weight of modern structures is carried by the steel or concrete frame, the walls do not support the building.

由于现代结构的重量由钢或混凝土框架承受,墙体不再支撑建筑物。

40. They have become curtain walls, which keep out the weather and let in light.

它们成为幕墙,将日晒风吹阻挡在外,同时让光线进入建筑。

41. In the earlier steel or concrete frame building, the curtain walls were generally made of masonry; they had the solid look of bearing walls.

在较早的钢或混凝土框架建筑中,幕墙一般由砌体构成;它们具有承重墙的结实外观。

42. Today, however, curtain walls are often made of lightweight materials such as glass, aluminum or plastic, in various combinations.

但是今天,幕墙通常由轻质材料组成,如玻璃、铝或塑料,并形成不同的组合。

43. Another advance in steel construction is the method of fastening together the beams.
钢结构中的另一个进步是梁的连接方式。

44. For many years the standard method was riveting.
在很多年里,连接的标准方式是铆接。

45. A rivet is a bolt with a head that looks like a blunt screw without threads.
铆钉是个有头的螺栓,看上去像个没有螺纹的圆头螺丝钉。

46. A rivet is heated, placed in holes through the pieces of steel, and a second head is formed at the other end by hammering it to hold it in place.
铆钉加热后穿过钢构件之间的孔洞,并通过锤击另一端而形成第二个铆钉头,从而将其固定就位。

47. Riveting has now largely been replaced by welding, the joining together of pieces of steel by melting a steel material between them under high heat.
如今铆接已大量地被焊接所替代,即钢构件间的连接通过高热熔化它们之间的钢材料(即焊条)进行。

48. Pre-stressed concrete is an improved form of reinforcement.
预应力混凝土是加强法的改进形式。

49. Steel rods are bent into the shapes to give them the necessary degree of tensile strength.
将钢筋弯成一定的形状以使它们具有必要的抗拉强度。

50. They are then used to prestress concrete, usually by one of two different methods.
然后用该钢筋对混凝土施加预应力,通常可采用两种不同方法中的任何一种。

Task Two   Sample Dialogue

Directions: *In this section, you are going to read several times the following sample dialogue about the relevant topic. Please pay special attention to five C's (culture, context, coherence, cohesion and critique) in the sample dialogue and get ready for a smooth communication in the coming task.*

In a Civil Engineering class

(*A teacher and his students are talking about the importance of learning Civil Engineering.*)

Wang: Let us share some knowledge about our major.

Bao: Great! I think that civil engineering professional is useful.

Wang: Yes. The so-called civil refers to all the water, soil, culture related to infrastructure construction plan, construction and repair.

Bao: From the narrow sense of definition, civil engineering is also called structural engineering, bridge and tunnel engineering, geotechnical engineering, highway and city road etc.

Wang: We mainly study the basic theory and knowledge of engineering mechanics,

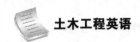

geotechnical engineering, structural engineering and water conservancy engineering.

Bao: We also accept the engineering drawing, engineering surveying, computer application, specialty experiments, structure design and construction practice of basic training.

Wang: At the same time, for the profession, we need to have some basic ability.

Bao: Yes, such as spatial imagination, application ability and strong practical ability.

Wang: And I think we also need to be careful, and we should develop meticulous style of work and hard-working spirit.

Bao: Yes, any discipline is rigorous, and it's true for civil engineering, as well.

Wang: Job prospects, policies of the state, economic development and the direction of civil engineering specialty are closely related.

Bao: The industry compensation is showing a tendency of higher technology management.

Wang: So, during the period of school we should lay a good foundation for our future work.

Bao: We will experience not only good mastery of professional theory and practice of knowledge, but also obtain the corresponding qualification certificate.

Wang: Yes, I believe God help those who help themselves and we will surely be paid on the harvest.

Bao: Right, thank you for listening.

Wang: Thank you!

Task Three   Simulation and Reproduction

Directions: *The class will be divided into three major groups, each of which will be assigned a topic. In each group, some students may be the teacher, while others may be students. In the process of discussion, please observe the principles of cooperation, politeness and choice of words. One of the groups will be chosen to demonstrate the discussion to the class.*

1) As a college student majoring in civil engineering, can you talk about the functions of your major briefly?

2) Why do you choose this major, civil engineering?

3) What impressed you most during your studying of civil engineering?

Task Four   Discussion and Debate

Directions: *The class will be divided into two groups. Please choose your stand in regard to the following controversy and support your opinions with scientific evidences. Please refer to the specialized terms and classical sentences in the previous parts of this unit.*

Someone says that architecture is not only the deposit but also the vehicle of human civilization. It embodies the ideologies, concepts, philosophies, religion totems and aesthetic values of all peoples across the world. But the others don't think so. What's your opinion? Can you interpret the style of Big Ben from those perspectives mentioned above?

V. After-class Exercises

1. *Match the English words in Column A with the Chinese meaning in Column B.*

A	B
1）soak	A）浸没
2）circuit	B）回路
3）sag	C）下垂
4）tension	D）润滑
5）durability	E）预冷
6）meld	F）持久性
7）lubricate	G）拉力
8）instill	H）浸泡
9）submerge	I）浸染
10）pre-cooling	J）混合组成

2. *Fill in the following blanks with the words or phrases in the word bank. Change the forms if it's necessary.*

circular	preserve	curtain walls	depth	retain
tensile strength	reinforce	expansion	sink	below

1) Curtain walls have become _____, which keep out the weather and let in light.

2) Steel rods are bent into the shapes to give them the necessary degree of _____.

3) Steel rods are embedded in concrete to make _____ concrete in concrete beams or structures where tension will develop.

4) They also complement each other in another way: they have almost the same rate of contraction and _____.

5) In the built-up parts of cities freeways generally are either depressed _____ ground level or elevated.

6) _____ walls are those walls subject to horizontal earth pressures due to the retention of earth behind them.

7) Another method of building an underwater tunnel is to _____ tubular sections into a trench dug at the bottom of a river or other body of water.

8) The use of terraced earth sheltered housing on steeply sloping hillsides can help _____ precious arable flat land in mountainous regions.

9) A circular shield has proved to be most efficient in resisting the pressure of soft ground, so most shield-driven tunnels are _____.

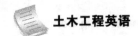

10）Shock waves also are utilized to determine the _____ of bedrock by measuring the time required for the shock wave to travel to the bedrock and return to the surface as a reflected wave.

3. *Translate the following sentences into English.*

    1）我们也接受工程制图、工程测量、计算机应用、专业试验、结构设计及施工实践等方面的基本训练。

    _____

    2）所谓的土木是指一切和水、土、文化有关的基础建设的计划、建造和维修。

    _____

    3）这个专业也需要我们具备一些基本的能力，例如空间想象力、应用能力以及较强的实践能力。

    _____

    4）在很多年里，连接的标准方式是铆接。

    _____

    5）事实上，他们协调工程中每个人的工作。

    _____

4. *Please write an essay of about 120 words on the topic: The Delicate Structure of The Big Ben. Some specific examples will be highly appreciated and watch out the spelling of some specialized terms you have learned in this unit.*

_____

_____

_____

_____

_____

_____

_____

_____

VI. Additional Reading

A Brief Introduction on the Elizabeth Tower

[A] Elizabeth Tower, previously called the Clock Tower but more popularly known as Big Ben, was raised as a part of Charles Barry's design for a new palace, after the old Palace of Westminster was largely destroyed by fire at the night of 16 October 1834. The new parliament was built in a neo-gothic style. Although Barry was the chief architect of the palace, he turned to Augustus Pugin for the design of the clock tower, which resembles earlier Pugin designs, including one for Scarisbrick Hall in Lancashire. The design for the tower was Pugin's last design before his final descent into madness and death, and Pugin himself wrote, at the time of Barry's last visit to him to collect the drawings: "I never worked so hard in my life for Mr. Barry for tomorrow I render all the designs for finishing his bell tower and it is beautiful."

Tower

Design

[B] The tower is designed in Pugin's celebrated Gothic Revival style, and the tower is 315 feet (96.0 m) high. The bottom 200 feet (61.0 m) of the tower's structure consists of brickwork with sand-colored Anston limestone(石灰岩) cladding(保护层). The remainder of the tower's height is a framed spire of cast iron. The tower is founded on a 50 feet (15.2 m) square raft, made of 10 feet (3.0 m) thick concrete, at a depth of 13 feet (4.0 m) below ground level. The four clock dials are 180 feet (54.9 m) above ground. The interior volume of the tower is 164, 200 cubic feet (4,650 cubic meters).

[C] Despite being one of the world's most famous tourist attractions, the interior of the tower is not open to overseas visitors, though United Kingdom residents were able to arrange tours (well in advance) through their Member of Parliament before the current repair works. However, the tower currently has no lift, though one is being installed, so those escorted had to climb the 334 limestone stairs to the top.

[D] Due to changes in ground conditions since construction, the tower leans slightly to the north-west, by roughly 230 millimeters (9.1 in) over 55 meters height, giving an inclination of approximately 1/240. This includes a planned maximum of 22 milimeters increased tilt(倾斜) due to tunneling for the Jubilee line extension. It leans by about 500 millimeters at the finial. Experts believe the tower's lean will not be a problem for another 4,000 to 10,000 years. Due to thermal(热量) effects it oscillates annually by a few millimeters east and west.

Name

[E] Journalists during Queen Victoria's reign called it St. Stephen's Tower. As MPs

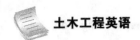

originally sat at St. Stephen's Hall, these journalists referred to anything related to the House of Commons asnews from "St. Stephens" (the Palace of Westminster contains a feature called St. Stephen's Tower, a smaller tower over the public entrance). The usage persists in Welsh, where the Westminster district, and Parliament by extension, is known as *San Steffan*.

[F] On 2 June 2012, *The Daily Telegraph* reported that 331 Members of Parliament, including senior members of all three main parties, supported a proposal to change the name from *Clock Tower to Elizabeth Tower* in tribute to Queen Elizabeth II in her diamond jubilee(周年庆祝) year. This was thought to be appropriate because the large west tower now known as Victoria Tower was renamed in tribute to Queen Victoria on her diamond jubilee. On 26 June 2012, the House of Commons confirmed that the name change could go ahead. The Prime Minister, David Cameron, announced the change of name on 12 September 2012 at the start of Prime Minister's Questions. The change was marked by a naming ceremony in which the Speaker of the House of Commons, John Bercow, unveiled a name plaque(匾牌) attached to the tower on the adjoining Speaker's Green.

Clock

Dials

[G] The clock and dials were designed by Augustus Pugin. The clock dials are set in an iron frame 23 feet (7.0 meter) in diameter, supporting 312 pieces of opal(猫眼石) glass, rather like a stained-glass window. Some of the glass pieces may be removed for inspection of the hands. The surround of the dials is gilded. At the base of each clock dial in gilt(镀金) letters is the Latin inscription:

DOMINE SALVAM FAC REGINAM NOSTRAM VICTORIAM PRIMAM

Which means *O Lord, keep safe our Queen Victoria the First.*

Unlike most other Roman numeral clock dials, which show the "4" position as "IIII", the Great Clock faces depict "4" as "IV".

Movement

[H] The clock's movement is famous for its reliability. The designers were the lawyer and amateur horologist(机械大师) Edmund Beckett Denison, and George Airy, the Astronomer Royal. Construction was entrusted to clockmaker Edward John Dent; after his death in 1853 his stepson Frederick Dent completed the work, in 1854. As the tower was not complete until 1859, Denison had time to experiment: instead of using the deadbeat esca pement(直进式擒纵机构) and remontoire(摆锤均衡键) as originally designed, Denison invented the double three-legged gravity escapement. This escapement provides the best separation between pendulum and clock mechanism. The pendulum(钟摆) is installed within an enclosed windproof box beneath the clock room. It is 13 feet (4.0 meter) long, weighs 660 pounds (300 kilogram), suspended on a strip of spring steel 1/64 inch in thickness, and beats every two seconds. The clockwork mechanism in a room below weighs five tons.

[I] On top of the pendulum is a small stack of old penny coins; these are to adjust the time of the clock. Adding a coin has the effect of minutely lifting the position of the pendulum's center of mass, reducing the effective length of the pendulum rod and hence increasing the rate at which the pendulum swings. Adding or removing a penny will change the clock's speed by 0.4 seconds per day. The clock is hand wound (taking about 1.5 hours) three times a week.

[J] On 10 May 1941, a German bombing raid damaged two of the clock's dials and sections of the tower's stepped roof and destroyed the House of Commons chamber. Architect Sir Giles Gilbert Scott designed a new five-floor block. Two floors are occupied by the current chamber, which was used for the first time on 26 October 1950. The clock ran accurately and chimed throughout the Blitz.

Bell

Great Bell

[K] The main bell, officially known as the Great Bell but better known as Big Ben, is the largest bell in the tower and part of the Great Clock of Westminster. It sounds an E-natural.

[L] The original bell was a 16 ton (16.3 tonne) hour bell, cast on 6 August 1856 in Stockton-On-Tees by John Warner & Sons. The bell was possibly named in honor of Sir Benjamin Hall, and his name is inscribed on it. However, another theory for the origin of the name is that the bell may have been named after a contemporary heavyweight boxer Benjamin Caunt. It is thought that the bell was originally to be called Victoria or Royal Victoria in honor of Queen Victoria, but that an MP suggested the nickname during a Parliamentary debate; the comment is not recorded in Hansard.

[M] Since the tower was not yet finished, the bell was mounted in New Palace Yard but, during testing it cracked beyond repair and a replacement had to be made. The bell was recast on 10 April 1858 at the Whitechapel Bell Foundry as a 131ton (13.76 tonne) bell. The second bell was transported from the foundry to the tower on a trolley drawn by sixteen horses, with crowds cheering its progress; it was then pulled 200 ft. (61.0 meter) up to the Clock Tower's belfry, a feat that took 18 hours. It is 7 feet 6 inches (2.29 meters) tall and 9 feet (2.74 meters) diameter. This new bell first chimed in July 1859; in September it too cracked under the hammer. According to the foundry's manager, George Mears, the horologist Denison had used a hammer more than twice the maximum weigh tspecified. For three years Big Ben was taken out of commission and the hours were struck on the lowest of the quarter bells until it was repaired. To make the repair, a square piece of metal was chipped out from the rim(边沿) around the crack, and the bell given an eighth of a turn so the new hammer struck in a different place. Big Ben has chimed with a slightly different tone ever since, and is still in use today with the crack unrepaired. Big Ben was the largest bell in the British Isles until "Great Paul", a 163

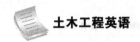

ton (17 tonne) bell currently hung in St Paul's Cathedral, was cast in 1881.

Chimes

[N] Along with the Great Bell, the belfry(钟楼)houses four quarter bells which play the *Westminster Quarters* on the quarter hours. The four quarter bells sound G, F, E, and B. They were cast by John Warner & Sons at their Crescent Foundry in 1857 (G, F and B) and 1858 (E). The Foundry was in Jewin Crescent, in what is now known as The Barbican, in the City of London. The bells are sounded by hammers pulled by cables coming from the link room—a low-ceiling space between the clock room and the belfry—where mechanisms translate the movement of the quarter train into the sounding of the individual bells.

[O] The quarter bells play a once-repeating, 20-note sequence of rounds and four changes in the key of E major: 1~4 at quarter past, 5~12 at half past, 13~20 and 1~4 at quarter to, and 5~20 on the hour (which sounds 25 seconds before the main bell tolls the hour). Because the low bell (B) is struck twice in quick succession, there is not enough time to pull a hammer back, and it is supplied with two wrench(猛扭) hammers on opposite sides of the bell. The tune is that of the Cambridge Chimes, first used for the chimes of Great St Mary's church, Cambridge, and supposedly a variation, attributed to William Crotch, based on violin phrases from the air "I know that my Redeemer liveth" in Handel's Messiah. The notional words of the chime, again derived from Great St Mary's and in turn an allusion to Psalm 37:23~24, are: "All through this hour/Lord be my guide/And by Thy power/No foot shall slide". They are written on a plaque on the wall of the clock room.

[P] One of the requirements for the clock was that the first stroke of the hour bell should be correct to within one second per day. The tolerance is with reference to Greenwich Mean Time (BST in summer). So, at twelve o'clock, for example, it is the first of the twelve hour-bell strikes that signifies the hour (the New Year on New Year's Eve at midnight). The time signaled by the last of the "six pips" (UTC) may be fractionally(很小) different.

(*If you want to find more information about this corporation, please log on https://en.wikipedia.org/wiki/Big_Ben*)

1. *Read the passage with ten sentences attached to it. Each statement contains information given in one of the paragraphs. Identify the paragraph from which the information is derived. You may choose a paragraph more than once. Each paragraph is marked with a letter.*

    1) Along with the Great Bell, the belfry houses four quarter bells which play the Westminster Quarters on the quarter hours.

    2) The clock's movement is famous for its reliability.

    3) Architect Sir Giles Gilbert Scott designed a new five-floor block.

    4) Elizabeth Tower was raised as a part of Charles Barry's design for a new palace, after the old Palace of Westminster was largely destroyed by fire on the night of 16

October 1834.

5) This was thought to be appropriate because the large west tower now known as Victoria Tower was renamed in tribute to Queen Victoria on her diamond jubilee.

6) Due to changes in ground conditions since construction, the tower leans slightly to the north-west.

7) So, at twelve o'clock, for example, it is the first of the twelve hour-bell strikes that signifies the hour (the New Year on New Years Eve at midnight).

8) The bell was possibly named in honor of Sir Benjamin Hall and his name is inscribed on it.

9) The usage persists in Welsh, where the Westminster district, and Parliament by extension, is known as San Steffan.

10) The clock dials are set in an iron frame 23 feet (7.0 m) in diameter, supporting 312 pieces of opal glass, rather like a stained-glass window.

1) _____ 2) _____ 3) _____ 4) _____ 5) _____
6) _____ 7) _____ 8) _____ 9) _____ 10) _____

2. *In this part, the students are required to make an oral presentation on either of the following topics.*

1) What are the characteristics of the Big Ben in terms of design?

2) What of the Big Ben impresses you most?

习题答案

Unit Eight   Victoria Tower

I. Pre-class Activity

Directions: *Please read the general introduction about the **Victoria Tower** and tell something more about the topic to your classmates.*

Victoria Tower

The Victoria Tower is the square at the south-west end of the Palace of Westminster in London, facing south and west onto black Rod's Garden and Old Place Yard. At 98.5 meters (323ft), it is slightly taller than the more famous Elizabeth Tower (formerly known as the clock Tower and popularly known as Big Ben) at the north end of the Palace (96.3 meters). It houses the Parliamentary Archives in archive conditions meeting the BS 5,454 standard, on 12 floors. All 14 floors of the building were originally linked via a single wrought-iron Victorian staircase of 553 steps, of which five floors survive.

The main entrance at the base of the tower is the Sovereign's Entrance, through which the Monarch passes at the State Opening of Parliament. On top of the Victoria Tower is an iron flagstaff from which flies the Union Flag, when the Sovereign is present in the Palace, the Royal Standard. The flag used to be flown only on days when either House Parliaments at, but since January 2010 it has been flown every day.

II. Specialized Terms

Directions: *Please memorize the following specialized terms before the class so that you will be able to better cope with the coming tasks.*

air-entrained concrete 加气混凝土

asphalt concrete 沥青混凝土

blinding concrete 垫层混凝土

bloating (bulking, expansion) agent 膨胀剂

bonding surface 结合面

bug hole (pitted surface) 麻面

byproduct n. 副产品

calcium particles 钙颗粒

cast-in-situ (site, place) concrete 现浇混凝土

cavity n. 洞,穴

clearance n. 清理

cold joint 冷缝

concrete pouring (placement, casting) 浇筑混凝土

contaminated material 污染物质

controlling parameter 控制参数

crack n. 裂缝

curing n. 养护

declination of the boundary 料界偏差

deformation n. 变形

fibrous concrete 纤维混凝土

fill placement record 填筑记录

fire-damaged concrete 火灾损伤混凝土

first stage concrete 第一层混凝土

fluidizer n. 塑化剂

foamed concrete 泡沫混凝土

formwork shifting (moving) 跑模

fog area 雾区

gas-forming admixture 引气剂

greencutting 绿化

grout leakage 漏浆

high-performance concrete 高性能混凝土

high-strength concrete 高强混凝土

honeycomb n. 蜂窝

incomplete vibration 漏振

irregularity adj. 不规则

lack of compaction 漏压

lean concrete (poor concrete) 贫混凝土

lens n. 透镜体

lightweight aggregate concrete 轻骨料混凝土

lump v. 结块

mass concrete 大体积混凝土

membrance curing 薄膜养护

moist curing 湿润养护

moisture condition 水分调节

muddy adj. 泥泞的

non-fines concrete 无砂混凝土

non-plastic concrete 干硬性混凝土

non-shrinkage concrete 无收缩混凝土

normal concrete 常态混凝土

normal curing 标准养护

oversize material 超粒径材料

place and spread 摊铺

plain concrete 素混凝土

plastic concrete 塑性混凝土

pocket n. 资金

porous concrete 多孔混凝土

pre-stressed concrete 预应力混凝土

pump concrete 泵送混凝土

reinforced concrete 钢筋混凝土

remove 清除

repairing 修补

replace 重新填筑

required embankment 必填方

retarding agent 缓凝剂

retarding water reducer 缓凝型减水剂

retreat (do the work again) 返工

roller compacted concrete (rollcrete) 碾压混凝土

scabbling n. 凿毛

scarify v. 翻松,刨毛

screen and wash 筛分和冲洗

second stage concrete 第二层混凝土

seepage n. 渗流

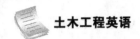

segregation n. 离析	trowel n. 泥刀，小铲子
separation n. 分离	trunk n. 树干
set-controlling admixture 调凝剂	truss n. 支撑屋顶、桥梁等的桁架
shear area 剪力区	tutorial n. 教程，使用说明书
stair(staggered joint) n. 楼梯口	utility n. 实用程序，实用品
steam curing 蒸汽养护	verification n. 确认
subzone 分区	vertex n. 顶点，最高点，头顶
super-plasticizer n. 塑化剂	vice n. 台钳，老虎钳
surface finishing 表面处理	withstand vt. 顶住，经受住
surface-active agent 表面活性剂	vibrating n. 振捣
take a sample 取样	water leakage 漏水
tie rod hole 拉杆孔	water seepage 渗水
trim v.修剪	water-reducing agent 减水剂

III. Watching and Listening

Task One　Victoria Tower（I）

New Words

视频链接及文本

decipher v.解读，破译	dominate adj. 主要的
soar v.耸立	precursor n. 前辈，先驱
ceremonial adj. 仪式的	act n. 法案
skeleton n. 骨骼；骨架	parchment scroll 羊皮纸卷轴
taper v.逐渐变细	humidity n. 湿度
girder n. 大梁	incredible adj. 难以置信的
inspiration n. 灵感	

Exercises

1. *Watch the video for the first time and choose the best answers to the following questions.*

　　1）It stands _____ feet high, even taller than Big Ben's clock tower.

　　　　A. 230　　　　　　　　　　　　　B. 320

　　　　C. 300　　　　　　　　　　　　　D. 200

　　2）The traditional bricks and stones hide a secret. An immense _____ skeleton.

　　　　A. wood　　　　　　　　　　　　B. iron

　　　　C. stone　　　　　　　　　　　　D. concrete

　　3）The _____ channel this immense load down to a base made from 13 tons girders.

　　　　A. columns　　　　　　　　　　　B. skeletons

　　　　C. walls　　　　　　　　　　　　D. beam

4) This tower was the inspiration for one of America's earliest _____.

    A. house   B. building

    C. skyscrapers   D. tower

5) Deep inside the Victoria Tower, holds _____'s most precious treasures just as it did when it was built.

    A. king   B. country

    C. government   D. parliament

2. *Watch the video again and decide whether the following statements are true or false.*

    1) Britain's Houses of Parliament, a building light years ahead of its time. (　)

    2) Yet at Houses of Parliament's base contains a vast, open, ceremonial entrance. (　)

    3) The Victoria Tower's engineering was ages ahead of its time. (　)

    4) Its concrete skeleton was a vital step towards developing skyscrapers in the 20th century. (　)

    5) Storing all of parliament's documents and laws push the Victoria Tower up to its record-breaking height. (　)

3. *Watch the video for the third time and fill in the following blanks.*

But why did they need to construct such an extraordinary structure? Deep inside the Victoria Tower, holds parliament's most precious _____ just as it did when it built. Around us, we can see 64,000 _____ of parliament. Dating back to the 15th century, _____ in parchment scrolls. One scroll _____ act. All the laws that have been _____ by parliament over the centuries. Archivist Caroline Shelton, _____ these priceless laws. It's an amazing space and _____, too. 16 degrees centigrade and 55% humidity which helps to keep all these incredible scrolls at the _____ temperature to _____ them for the future. Storing all of parliament's documents and laws _____ the Victoria Tower up to its record-breaking height.

4. *Share your opinions with your partners on the following topic for discussion.*

    1) Do you know anything about the Victoria Tower? What is the vital step in constructing this structure?

    2) Can you use a few lines to list your understanding about steel-frame structure?

Task Two　Victoria Tower (II)

New Words

scaffolding n. 脚手架	investigate v.调查
crane n. 吊车,起重机	available adj. 可用的
hoist v.升起,吊起	tidal adj. 潮汐的
crank v.装曲柄	timber pile 木桩
gear n. 齿轮	trailblazing adj. 开拓性的
adversary n. 对手	timber pile 木桩

视频链接及文本

137

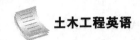

barrier n. 障碍物

watertight adj. 不漏水的

cofferdam n. 围堰

drain v.排水,流水

withstand v.抵挡

mighty adj. 有力的

permanent adj. 永久的

massive adj. 大量的

taper v.逐渐减少

Exercises

1. *Watch the video for the first time and choose the best answers to the following questions.*

  1) There was no space for the _____ or cranes that could create such a vast tower.

    A. construction   B. wall

    C. scaffolding   D. platform

  2) On top, a _____ lifted the 4 tons of stones from the ground to the platform.

    A. scaffolding   B. gear

    C. crane   D. man

  3) To observers, the tower appeared to grow _____ under its own steam.

    A. heavily   B. effortlessly

    C. slowly   D. quickly

  4) At parliaments river wall, where the Houses of Parliament meets the river _____, structural engineer Alistair Silencer investigates.

    A. Yellow   B. Yangtze

    C. Amazon   D. Thames

  5) So we can see that all this part of the building had to be built _____.

    A. out into the existing tidal River   B. back off the existing tidal River

    C. out into the former tidal River   D. back off the former tidal River

2. *Watch the video again and decide whether the following statements are true or false.*

  1) Two more cranes placed the heavy stones in position below the tower walls. (　　)

  2) But on the other side, they had to contend with an even greater adversary. The River Thames. (　　)

  3) To build on this land, Victorians engineers needed to avoid the river. (　　)

  4) First, they took twin rows of timber piles, each 63 feet deep, and drove them into the riverbed. (　　)

  5) The engineers filled the space with stone to create a watertight wall called a cofferdam. (　　)

3. *Watch the video for the third time and fill in the following blanks of the following paragraph.*

  The final challenge was to _____ the temporary cofferdam with something strong enough to withstand the _____ Thames. So once the cofferdam had been _____ and they would create the dry space behind it. That allowed them to start construction of the _____ river wall. This wall needs to protect parliament, from the extreme _____ of the rising tide. The secret to the strength of the river wall, lies in its _____ . At high tide the pressure _____ down, the river wall itself needs to get wider and more massive to stay _____ under the water pressure. Behind me you can see the shape, which tapers _____ . Today, dams all over the world, use this same taper design. It has kept the Houses of parliament standing _____ in the river for over 160 years.

4. *Share your opinions with your partners on the following topic for discussion.*

1）Do you know Britain's Houses of Parliament? At its southern end soars the extraordinary Victoria Tower. How did engineers build this 320-foot-high mega structure and support its nine floors above the empty space?

2）On the other side of the Victoria Tower, the Victorian engineers had to contend with an even greater adversary: the River Thames. What they have done to deal with this river?

IV. Talking

Task One   Classical Sentences

Directions: *In this section, some popular sentences are supplied for you to read and to memorize. Then, you are required to simulate and produce your own sentences with reference to the structure.*

General Sentences

1. Children enter school at the age of five, don't they?
   孩子们到五岁时就上学,是吗?

2. In elementary school, the child learns to read and write.
   在小学,孩子们学习读和写。

3. In secondary school, children get more advanced knowledge.
   在中学,孩子们学到更高难度的知识。

4. In universities, students are trained to become teachers and engineers.
   在大学里,学生们被培养成老师和工程师。

5. He went to grade school in New York and high school in Chicago.
   他在纽约上小学,在芝加哥上中学。

6. In college I majored in science. What was your major?
   大学里我的专业是科学。你呢?

7. My sister graduated from high school. Graduation was last night.
   我姐姐中学毕业了。昨晚举行了毕业晚会。

8. I'm a graduate of Yale University. I have a Bachelor of Arts degree.
   我是一名耶鲁大学的毕业生。我获得了艺术学士学位。

9. If you expect to enter the university, you should apply now.

如果你想上大学,你现在就应该申请。

10. This is my first year of college. I'm a freshman.

这是我大学的第一年,我是大一新生。

11. My uncle is a high school principal.

我叔叔是一名中学校长。

12. What kind of grades did you make in college?

你在大学中成绩怎么样?

13. During your first year of college, did you make straight A?

你大学一年级时,成绩全优吗?

14. My brother is a member of the faculty. He teaches economics.

我的哥哥是一名老师,他教经济学。

15. John has many extracurricular activities. He's in the football team.

约翰有许多课外活动。他是一名足球队员。

16. I'm a federal employee. I work for the Department of Labor.

我是一名联邦雇员。我在劳工部工作。

17. What kind of work do you do? Are you a salesman?

你做什么工作? 是不是销售员?

18. As soon as I complete my training, I'm going to be a bank teller.

一旦我完成了我的培训,我将成为一名银行出纳员。

19. John has built up his own business. He owns a hotel.

约翰已经有了自己的生意,他有一个旅馆。

20. What do you want to be when you grow up?

你长大后想做什么?

21. My son wants to be a policeman when he grows up.

我儿子长大后想当一名警察。

22. I like painting, but I wouldn't want it to be my life's work.

我喜欢绘画,但是我不会让绘画成为我终生的职业。

23. Have you ever thought about a career in the medical profession?

你是否考虑过成为一名医药行业的从业人员?

24. My uncle was a pilot with the airlines. He has just retired.

我的叔叔是航空公司的飞行员,他刚刚退休。

25. My brother's in the army. He was just promoted to the rank of major.

我的哥哥在军队,他刚刚被提升为少校。

26. I have a good-paying job with excellent hours.

我有一份工资很高、工作时间理想的工作。

27. My sister worked as a secretary before she got married.

我的姐姐结婚前是个秘书。

28. George's father is an attorney. He has his own company.
    乔治的父亲是个律师,他拥有自己的公司。

29. He always takes pride in his work. He's very efficient.
    他总是以他的工作为荣。他是个很能干的人。

30. Mr. Smith is a politician. He's running for election as governor.
    史密斯先生是个政治家。他正在为竞选州长而奔忙。

31. After a successful career in business, he was appointed ambassador.
    他在生意中有了成就后,就被任命为企业代表。

32. Why is Mr. Smith so tired? Do you have any idea?
    为什么史密斯先生这么累? 你知道吗?

33. According to Mr. Green, this is a complicated problem.
    听格林先生说,这是一个复杂的问题。

34. I wish you would give me a more detailed description of your trip.
    我希望你能更详细地描述一下你的旅行。

35. We used to have a lot of fun when we were at that age.
    我们那么大时经常玩得很开心。

36. I never realized that someday I would be living in New York.
    我从没有想过,有一天我能住在纽约。

37. We never imagined that John would become a doctor.
    我们从来没有想过约翰会成为一名医生。

38. I beg your pardon. Is this seat taken?
    请问,这个座位有人坐吗?

39. The waiter seems to be in a hurry to take our order.
    服务员似乎急着要我们点东西。

40. —Which would you rather have, steak or fish?
    —I want my steak well-done.
    —你想来点什么,牛排还是鱼?
    —我想要全熟的牛排。

41. —What kinds of vegetables do you have?
    —I'll have mashed potatoes.
    —你想要什么蔬菜?
    —我想要些土豆泥。

42. —What do you want?
    —I want a cup of coffee.
    —你想要点什么?
    —我想来杯咖啡。

43. Which one would you like, this one or that one?
    你想要哪一个? 这一个还是那一个?

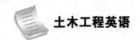

44. It doesn't matter to me.

随便都可以。

45. Would you please pass the salt?

麻烦把盐递过来好吗?

46. They serve good food in this restaurant.

这家餐馆的东西很好吃。

47. Are you ready for your dessert now?

现在可以吃点心了吗?

48. This knife/fork/spoon is dirty. Would you bring me a clean one, please?

这把刀/叉/勺脏了,麻烦你拿一个干净的来好吗?

49. You have your choice of three flavors of ice cream. We have vanilla, chocolate and strawberry.

你可以选择三种口味的冰淇淋,我们有香草味的,巧克力味的,草莓味的。

50. The restaurant was filled, so we decided to go elsewhere.

这家餐馆已经没餐位了,我们得去别处了。

Specialized Sentences

1. Construction planning is a fundamental and challenging activity in the management and execution of construction projects.

在施工项目的管理和执行中,施工计划是一项基本的并且具有挑战性的活动。

2. It involves the choice of technology, the definition of work tasks, the estimation of the required resources and duration for individual tasks, and the identification of any interactions among the different work tasks.

它涉及技术的选择、工作任务的定义、每一个任务所需资源、时间的估算,以及不同工作任务之间相互作用的识别等工作。

3. A good construction plan is the basis for developing the budget and the schedule for work.

一个好的施工计划是制定预算和工作进度的基础。

4. For example, the extent which sub-construction will be used on a project is often determined during construction planning.

例如,分包商在一个项目中所承担的范围通常是在计划期间决定的。

5. Developing the construction plan is a critical task in the management of construction, even if the plan is not written or otherwise formally recorded.

确定施工计划是管理当中的一项重要工作,即使这个施工计划没有采取书面的或其他正式的形式确定下来。

6. Some projects are primarily divided into expense categories with associated costs.

一些项目被大致分为费用类别及其相关成本。

7. In these cases, construction planning is cost or expense oriented.

在这些情况下,施工计划是成本或费用导向的。

8. For example, borrowing expenses for project financing and construction planning overhead items are commonly treated as indirect costs.

例如,为项目融资而产生的借款费用和日常开支项目通常被认为是间接成本。

9. For other projects, scheduling of work activities over time is critical and is emphasized in the planning process.

对其他项目来说,制定工作的时间的进度非常重要,而且也会在计划过程中受到重视。

10. In this case, the planner insures that the proper precenes among activities are maintained and that efficient scheduling of the available resources prevails.

在这种情况下,施工计划人员应当确保维持各项工作适当的前导关系,保证对各项已有资源的合理计划。

11. All of the civil work on the field will be executed by us (our company).

所有现场的土建工作都将由我们(我公司)承担实施。

12. Civil engineering design is performed on our own at home.

土建工程是根据卖方提供的技术条件设计的。

13. This is a working (plot plan, vertical layout, structure plan, floor plan, general plan) drawing.

这是施工(平面布置、竖向布置、结构、屋面、总)图。

14. The quality of civil engineering conforms with our domestic technical standard (China National Building Code).

土建工程的质量符合我们国内的技术标准(中国工程建设标准规范)。

15. Our civil work include construction of roads, buildings, foundations and reinforced concrete structure.

我们的土建工作包括建造道路、建筑物、基础和钢筋混凝土结构。

16. It will take two weeks to complete this building (foundation).

这房子(基础)需要两周时间才能完成。

17. Sand, brick and stone are generally used in constructing houses.

砂、砖和石头通常用于建造房屋。

18. These are the anchor bolts (rivets, unfinished bolts, high-strength structural bolts) for the structure.

这是用于结构的锚定螺栓(铆钉、粗制螺栓、高强度结构螺栓)。

19. The holes of anchor bolts will be grouted with normal (Portland, non-shrinkage) cement mortar.

这些地脚螺孔将灌入普通(波特兰、无收缩)水泥砂浆。

20. We usually measure the strength of concrete at 28 days after which has been cast.

我们通常在混凝土浇筑后 28 天测定其强度。

21. The average compressive strength of samples is $500kg/cm^2$.

试样的平均抗压强度为 500 千克/平方厘米。

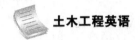

22. Our concrete material is mixed in a rotating-drum batch mixer at the job site.

我们用的混凝土是在现场的间歇式转筒搅拌机中搅拌的。

23. Quality of concrete depends on proper placing, finishing and curing.

混凝土的质量取决于正确的浇筑、抹面和养护。

24. The concrete can be made stronger by pre-stressing in our factory.

在我们厂里可以使混凝土通过施加预加应力得到增强。

25. Most of construction material can be tested in our laboratory.

我们的试验室可以检验大部分建筑材料。

26. We shall finish the civil work by the end of the year.

在年底前我们将完成土建工作。

27. We have all kinds of construction machinery on the job site.

我们在现场有各种施工机械。

28. The truck crane (gantry crane, mobile slewing crane, bridge crane, crawler crane) can lift a weight of 15 tons.

这台汽车起重机(龙门吊、悬臂汽车吊、桥式吊、履带式起重机)能吊起 15 吨的重物。

29. The hoisting capacity of that gin pole (girder pole, guy derrick) is sixty tons.

那个起重抱杆(格状抱杆、转盘抱杆)的起重能力为 60 吨。

30. The brand of this truck is JIEFANG (YUEJIN, HUANGHE, HINO, NISSAN, TOYOTA, ISUZU, TADANO, KATO, MITSUBISHI, HITACHI, IFA, BENZ, SKODA, TATRA, CSEPEL, FIAT, FORD, DODGE, GMC, BERALIN, ROMAN, ГИС,ГИП).

这台载重汽车的品牌是"解放"(跃进、黄河、日野、日产、丰田、五十铃、多田野、加藤、三菱、日立、依发、奔驰、斯柯达、太脱拉、却贝尔、菲亚特、福特、道奇、通用、贝埃标、罗曼、吉斯、吉尔)。

31. This is a motor hoist (winch) with low (high) wrapping speed.

这是低(高)速电动卷扬机。

32. We used to weld pipes with direct current (D.C) welder (alternating current A.C. welder).

我们总是用直流(交流)电弧焊机焊接管子。

33. The concrete mixer (concrete truck mixer, concrete vibrator, concrete batch plant) made by Huadong Works are steady in quality and reliable in performance.

华东工厂生产的混凝土搅拌机(搅拌车、混凝土振动器、混凝土搅拌站)质量稳定、性能可靠。

34. Loaders, fork lifts and air compressors are heavy construction machinery.

装载机、叉车和空气压缩机都是重型施工机械。

35. We must study the instruction before operation the pipe bender (hand pump, hydraulic testing pump).

在操作弯管机(手压泵、液压试压泵)之前,我们必须先仔细阅读说明书。

36. The thickness of the steel plate handled by this roller mill (shear machine) is 25 millimeters.

这台滚压机(剪切机)加工的钢板厚度可达25毫米。

37. This lathe (milling, boring, grinding, drilling, gear cutting, and planer) machine is of home manufacture.

这台车床(铣床、镗床、磨床、钻床、切齿机、刨床)是我们国内制造的。

38. All machinery must be lubricated periodically according to the lubrication chart.

所有的机械都必须根据润滑图表定期加油。

39. To maintain its efficiency, the machinery needs a regular service (check-over, repairing, overhaul).

这台机械需要进行一次定期保养(全面检查、修理、大修),以维持其工作效率。

40. This machinery is made in China (Great Britain, USA, Canada, Italy, Holland, Japan, France, Romania, Bulgaria, Poland, Czechoslovakia, Hungary, Swiss, Sweden).

这台机械是中国(英国、美国、加拿大、意大利、荷兰、日本、法国、罗马尼亚、保加利亚、波兰、捷克斯洛伐克、匈牙利、瑞士、瑞典)制造的。

41. Tell me about the operation of the machine.

请告诉我如何操作这台机器。

42. Bearings must be lubricated periodically.

轴承必须定期润滑。

43. Need we make any adjustment of this machine?

我们有必要调整这台机器吗?

44. There are many tools in my tool storage unit (tool chest, tool box).

我的工具柜(工具盒、工具箱)里面有很多工具。

45. Get me a hammer (hacksaw, file, scraper, chisel, socket, wrench, hook spanner, adjustable wrench, pipe wrench, ratchet wrench, open end wrench, screw driver, hand vice, pliers, pocket knife).

给我拿一把手锤(钢锯、锉刀、刮刀、凿子、套筒扳手、钩扳手、活动扳手、管扳手、棘轮扳手、开口扳手、螺丝刀、手钳、扁嘴钳、小刀)。

46. Straightedge rule (square rule, slide gauge, inside and outside micrometer, steel tape, feeler, dial gauge, depth micrometer, wire gage, radius gage, thread pitch gage) is a kind of common measuring tool.

直尺(角尺、游标卡尺、内径和外径千分尺、钢卷尺、塞尺、千分表、深度千分尺、线规、半径规、螺距规)是一种常用量具。

47. The precision of this fitter level (cross-test level) is 0.02mm/m.

这个钳工水平仪(框架式水平仪)的精度为0.02毫米/米。

48. We have got the instrument (pressure gauge, thermometer, tachometer, current meter, universal meter) ready for the experiment (test).

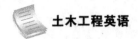
我们已经准备好做实验(试验)的仪器(压力表、温度计、转速计、电流表、万用表)。

49. That is an air (electric) powered grinder (portable grinder, angle grinder, straight grinder, drill, impact wrench, riveting hammer, hammer drill).

那是一个气(电)动砂轮机(手持砂轮机、角型砂轮、直型砂轮、钻机、冲击扳手、铆钉锤、锤钻机)。

50. Our electrical tools are double insulated and approved to international safety standards.

我们的电动工具都是双重绝缘的,并符合国际安全标准。

Task Two   Sample Dialogue

Directions: *In this section, you are going to read several times the following sample dialogue about the relevant topic. Please pay special attention to five C's (culture, context, coherence, cohesion and critique) in the sample dialogue and get ready for a smooth communication in the coming task.*

The Owner Hands over the Site Late

(*Mr. Bian, Project Manager from the Contractor, and Mr. Cheng, Chief Engineer, are talking to the Owner's Representative, Mr. Green, who looks very young and energetic, and the Consultant, Mr. Shaw, who seems to be quite experienced.*)

Bian: Last week, we sent you a letter, asking you to hand over the whole site to us, including the access road. But we haven't received a reply yet. The Contract is quite specific about this in its special conditions: Possession of the Site shall be given to the Contractor on the date named in the Appendix 1, which is April 20, 2008. Today is April 26. It's already one week late.

Green: I'm sorry about this, Mr. Bian. Unfortunately we have met with some difficulty in requisitioning the land on the left bank of the river for the site areas. You know, some of the land is privately owned and the owners won't agree to sell the land. Nor do they want to grant us permission to use it because they are afraid that the project will disturb their peaceful life.

Shaw: Some people here are obsessed with their traditional life. They don't want to have a change of their life! You'll never understand them.

Cheng: It's so puzzling! They should know that they will get benefits from this project. At least they could get the lighting power easily, at a much lower price.

Green: We've already promised to provide them with electric power for lighting free of charge. In return, they let us use their land for nothing.

Bian: When do you think you can solve the problem and make the whole site and the access road available to us? This is what we are concerned about.

146

Shaw: At present, the access road has been built up to the dam site and we are beginning to build the road to the power-house from the junction. It happens to be the rainy season and the rains have slowed down our progress, but we are making a great effort to finish it soon.

Green: So we can hand over to you the completed access road and the area on the right bank so that you can begin the preparation work there. You can have them from tomorrow. I'll give you a letter of confirmation right after this meeting. Once we solve our problem with the landowners, which I believe we can soon, we will hand over the remaining part of the site area.

Bian: Mr. Green, let me make it plain to you. The delay in handing over the site area has adversely affected our construction plan. We have to request you to extend the completion time of this project accordingly. Meanwhile, we reserve the right to be reimbursed for any costs incurred because of the delay. Fifteen engineers and technicians are already here waiting eagerly to start their jobs. A lot of preparation work needs to be done by us, especially the traffic road within the site.

Green: I understand your position, Mr. Bian. As for the compensation, we'll try to settle this matter according to the contract.

Bian: To make this project a success, we need cooperation from each other.

Green: I couldn't agree more.

Task Three   Simulation and Reproduction

Directions: *The class will be divided into two major groups, each of which will be assigned a topic. In each group, some students may be the teacher, while others may be students. In the process of discussion, please observe the principles of cooperation, politeness and choice of words. One of the groups will be chosen to demonstrate the discussion to the class.*

1) Suppose you are a chief engineer of your company. How to resolve the dispute about construction planning with the project manager?

2) Do you know any famous tower in China? Describe the construction features of the tower.

Task Four   Discussion and Debate

Directions: *The class will be divided into two groups. Please choose your stand in regard to the following controversy and support your opinions with scientific evidences. Please refer to the specialized terms and classical sentences in the previous parts of this unit.*

Yellow Crane Tower, located on Snake Hill in Wuchang, is one of the "Three Famous Towers South of Yangtze River (the other two: Yueyang Tower in Hunan and Tengwang Tower in Jiangxi)". Some people believe that the Yellow Crane Tower is the representative of the

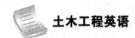

Chinese tower structure, but the others don't think so. They say that the Yellow Crane Tower is just very famous in south China. While in north China, there are many famous towers such as Fei Hong tower in Shanxi which can be the representative of the Chinese tower structure. What's your opinion?

V. After-class Exercises

1. *Match the English words in Column A with the Chinese meaning in Column B.*

A	B
1) scarify	A) 收面
2) contaminated material	B) 沥青混凝土
3) seepage	C) 翻松, 刨毛
4) controlling parameter	D) 速凝剂
5) asphalt concrete	E) 跑模
6) foamed concrete	F) 污染物质
7) pre-stressed concrete	G) 控制参数
8) surface finishing	H) 泡沫混凝土
9) fram-work shifting	I) 预应力混凝土
10) accelerator	J) 渗流

2. *Fill in the following blanks with the words or phrases in the word bank. Change the forms if it's necessary.*

strength	plan	heavy	manage	precision
cost	safety	execute	quality	lubricate

1) A good construction _____ is the basis for developing the budget and the schedule for work.

2) Construction planning is a fundamental and challenging activity in the _____ and execution of construction projects.

3) Some projects are primarily divided into expense categories with associated _____.

4) All of the civil work on the field will be _____ by us (our company).

5) The _____ of civil engineering conforms with our domestic technical standard (China National Building Code).

6) We usually measure the _____ of concrete at 28 days after which has been cast.

7) Loaders, fork lifts and air compressors are _____ construction machinery.

8) All machinery must be _____ periodically according to the lubrication chart.

9) The _____ of this fitter level (cross-test level) is 0.02mm/m.

10) Our electrical tools are double insulated and approved to international _____ standards.

3. *Translate the following sentences into English.*

    1) 在这些情况下,施工计划是成本或费用导向的。

    _____

    2) 土建工程是根据卖方提供的技术条件设计的。

    _____

    3) 这些地脚螺孔将灌入普通(波特兰,无收缩)水泥砂浆。

    _____

    4) 这是低(高)速电动卷扬机。

    _____

    5) 轴承必须定期润滑。

    _____

4. *Please write an essay of about 120 words on the topic: The beauty of tower. Some specific examples will be highly appreciated and watch out the spelling of some specialized terms you have learned in this unit.*

_____

_____

_____

_____

_____

_____

_____

_____

_____

_____

_____

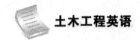

VI. Additional Reading

Victoria Tower

[A] The Victoria Tower is the tallest tower in the Palace of Westminster. Named after Queen Victoria, it was for many years the tallest and largest stone square tower in the world, with a height of 98.5 metres (325 feet). The tower was originally designed as a royal entrance and a repository(贮藏室) for the records of Parliament, and is now home to the Parliamentary Archives. On top of the tower is an iron flagstaff(旗杆). From here either the Royal Standard (if the Sovereign is present in the Palace) or the Union flag is flown.

Statues

[B] Due to the tower's prominent position and its part in royal ceremony, its architect Charles Barry designed particularly rich carving and sculpture for its interior and the underside of the entrance arch. These include statues of the patron(赞助人) saints of England, Scotland, Ireland and Wales, a life-size statue of Queen Victoria and two allegorical(寓言的) figures of Justice and Mercy.

[C] The gateways of the tower were built wide enough to allow the Queen's Coach to drive through for State Openings of Parliament. As Queen Victoria was the reigning monarch during the reconstruction of the Palace, the monogram(字母组合图案) VR appears throughout the Palace as do numerous other royal emblems, such as the Tudor rose and Portcullis. The restoration of the Victoria Tower between 1990 and 1994 required 68 miles of scaffolding tube, and one of the largest independent scaffolds in Europe. Some 1,000 cubic feet of decayed(腐烂) stonework was replaced, and over 100 shields were re-carved on site by a team of stonemasons.

[D] At the base of the tower is the Sovereign's Entrance, which is used by the Queen whenever entering the Palace. The steps leading from there to the Norman Porch are known as the Royal Staircase and are the start of the processional route taken by the Queen. By tradition, this route is the only one the Sovereign is allowed to take when he or she comes to the House of Lords. The Norman Porch is so called, because it was originally intended to house statues of the Norman kings.

[E] The new Palace of Westminster was custom-built by the Victorian architect Charles Barry for Parliamentary use. The design and layout of the building were thus carefully designed to serve the needs and workings of Parliament. In particular, Barry placed the location of the Sovereign's throne, the Lords Chamber and the Commons Chamber in a straight line, thus linking the three elements of Parliament in continuous form.

Design and detail

[F] Barry was also careful to weld the old to the new, so that the surviving medieval buildings—Westminster Hall, the Cloisters and Chapter House of St Stephen's, the Undercroft Chapel—formed an integral part of the whole. In his design, Barry was also concerned to balance the horizontal (which he emphasised with continuous bands of panelling) with the vertical (which he marked with turrets that ended high above the walls). He also introduced steeply-pitched iron roofs which emphasised the Palace's lively skyline. His Gothic scheme for the new Palace also extended to its interior furnishings, such as wallpapers, carvings, stained glass and even the royal thrones and canopies.

[G] Towering over the modest brick-built Georgian terraces of Westminster, the new Palace had an enormous effect on the imagination of the Victorian public. It also had a significant influence on the subsequent design of various public buildings such as town halls, law courts and schools throughout the country, and internationally.

Parliament on fire in 1834

[H] In 1834, the Exchequer was faced with the problem of disposing two cart-loads of wooden tally sticks. These were remnants(剩余部分) of an obsolete accounting system that had not been used since 1826. When asked to burn them, the Clerk of Works thought that the two underfloor stoves in the basement of the House of Lords would be a safe and proper place to do so.

[I] On 16 October, a couple of workmen arrived in the morning to carry out his instructions. During the afternoon, a party of visitors to the House of Lords, conducted by the deputy housekeeper Mrs Wright, became puzzled by the heat of the floor, and by the smoke seeping through it. But the workmen insisted on finishing their job. The furnaces were put out by 5pm, and Mrs Wright, no longer worried, locked up the premises. At 6 p.m., Mrs Wright heard the terrified wife of a doorkeeper screaming that the House of Lords was on fire. In no time, the flames had spread to the rest of the Palace. It was a great sight for the crowds on the streets (who were kept back by soldiers) and a great opportunity for artists such as J.M.W. Turner who painted several canvases(帆布) depicting it.

[J] Both Houses of Parliament were destroyed along with most of the other buildings on the site. Westminster Hall was saved largely due to heroic fire fighting efforts, and a change in the direction of the wind during the night. The only other parts of the Palace to survive were the Jewel Tower, the Undercroft Chapel, the Cloisters and Chapter House of St Stephen's and Westminster Hall.

Rebuilding the Palace

[K] The present-day Palace of Westminster is built in the perpendicular(垂直的) Gothic style, which was popular during the 15th century and was responsible for the Gothic revival of the 19th century. In 1835, a Royal Commission was appointed to study the rebuilding of the Palace and a heated public debate over the proposed styles ensued.

[L] The neo-classical style, similar to that of the White House in the United States, was popular at that time. However, as the design was associated with revolution and republicanism while the Gothic style was felt to embody conservative values, the commission announced in June 1835 that the style of the buildings should either be Gothic or Elizabethan. The commissioners also decided not to retain the original layout of the old palace, although the new design should incorporate the surviving Westminster Hall, the Undercroft Chapel and the Cloisters of St Stephen's.

[M] In 1836, the commissioners organised a public competition to design a new Palace in either of these styles. They received 97 entries, each identifiable only by a pseudonym or symbol. From these, the commissioners chose four, of which they were unanimous in preferring entry number 64 which bore the emblem of the Portcullis. This was the entry submitted by Charles Barry, who had proposed a Gothic-styled palace in harmony with the surviving buildings.

[N] The construction of the new Palace began in 1840. While Barry estimated a construction time of six years, at an estimated cost of £724,986, the project in fact took more than 30 years, at a cost of over £2 million. The first stone of the building was laid by Barry's wife on 27 August 1840. The site was extended into the river by reclaiming land, to a total of about eight acres. While construction was underway the House of Commons temporarily sat in the repaired Lesser Hall and the House of Lords used the Painted Chamber.

The House of Lords first sat in their new purpose-built chamber in 1847 and the House of Commons in 1852 (at which point Charles Barry received a knighthood). Although much of the rest of the building was completed by 1860, construction was not finished until a decade afterwards.

1. *Read the passage with ten sentences attached to it. Each statement contains information given in one of the paragraphs. Identify the paragraph from which the information is derived. You may choose a paragraph more than once. Each paragraph is marked with a letter.*

   1) He also introduced steeply-pitched iron roofs which emphasised the Palace's lively skyline.
   2) In 1834, the Exchequer was faced with the problem of disposing two cart-loads of wooden tally sticks.
   3) The tower was originally designed as a royal entrance and a repository for the records of Parliament, and is now home to the Parliamentary Archives.
   4) The gateways of the tower were built wide enough to allow the Queen's Coach to drive through for State Openings of Parliament.
   5) Both Houses of Parliament were destroyed along with most of the other buildings on the

site.

6) By tradition, this route is the only one the Sovereign is allowed to take when he or she comes to the House of Lords.

7) These include statues of the patron saints of England, Scotland, Ireland and Wales, a life-size statue of Queen Victoria and two allegorical figures of Justice and Mercy.

8) The design and layout of the building were thus carefully designed to serve the needs and workings of Parliament.

9) Towering over the modest brick-built Georgian terraces of Westminster, the new Palace had an enormous effect on the imagination of the Victorian public.

10) The site was extended into the river by reclaiming land, to a total of about eight acres.

1) _____ 2) _____ 3) _____ 4) _____ 5) _____

6) _____ 7) _____ 8) _____ 9) _____ 10) _____

2. *In this part, the students are required to make an oral presentation on either of the following topics.*

1) The special design of the towers which impressed you most.

2) Retell the construction structure of the Victoria Tower.

习题答案

Unit Nine Tunnel Construction

I. Pre-class Activity

Directions: *Please read the general introduction about* **Tunnel Construction** *and tell something more about the great architecture to your classmates.*

Tunnel Construction

Tunnels are dug in types of materials varying from soft clay to hard rock. The method of

tunnel construction depends on such factors as the ground conditions, the ground water conditions, the length and diameter of the tunnel drive, the depth of the tunnel, the logistics of supporting the tunnel excavation, the final use and shape of the tunnel and appropriate risk management.

There are three basic types of tunnel construction in common use:

1) Cut-and-cover tunnel, constructed in a shallow trench and then covered over.

2) Bored tunnel, constructed in situ, without removing the ground above. They are usually of circular or horseshoe cross-section. Some concepts of underground mining apply. Modern techniques include Shotcrete used in the New Austrian tunnelling method, use of a tunnel boring machine or tunnelling shield. But still tunnels are constructed which are secured with pit props and shoring and then are stained or timer supports are set. Techniques known from barrel vaults are helpful.

3) Immersed tube tunnel, sunk into a body of water and laid on or buried just under its bed.

II. Specialized Terms

Directions：*Please memorize the following specialized terms before the class so that you will be able to better cope with the coming tasks.*

acceptance n. 验收

age n. 龄期

air supply 供气

air test 通风试验

air-pocket 气囊

airtightness n. 气密性

block n. 仓号

bond breaker 脱模剂

cement anchor n. 锚固卷

circuit n. 回路

circuit and elements 管路及组件

coarse aggregate 粗骨料

coil n. 蛇形管

compressive stress 压应力

concrete class 混凝土等级

concrete plug 混凝土封堵

concrete precast element 混凝土预制件

contamination pollution 污染

conversion n. 转换

cylinder n. 缸套

double nipple 对丝连接套

epoxy mortar application 环氧砂浆抹面

erecting (setting up, fixing) formwork 立模

expansion coupling 伸缩节

female-quick coupling 内螺纹快接头

flat formwork 平面模板

flushing test 通水试验

formwork oil 模板油

gabion n. 铁丝笼

gap grouting 接缝灌浆

graduation n. 级配,分等级

graduation curve 级配曲线

gravity grouting 自重灌浆

horizontal bar 水平筋

humidity n. 湿度

identify v.识别,标识

impermeability 抗渗性

ingredient n. 成本,配料

initial setting strength 初凝强度

inspection n. 视查,检查,验收

intact adj. 未被触动的;完整的

inter-tower joint 塔间缝

introduce into 将穿入

isolate v.隔离

jet grouting 旋喷灌浆

joint washing 洗缝

key groove 楔形槽

laitance n. 浮浆皮

lap length 搭接长度

layer height 层高

main stress bar 主应力筋

meet design requirement 满足设计要求

mix design 配比设计

moment of inertia 惯性矩

multi-stage grouting 多次灌浆

overlap v.搭接

pigtail anchor 尾纤锚

pocket (box out, preset hole)预留孔

post tensioning 后张拉

post-cooling 后冷却

pre-cooling 预冷却

preliminary inspection 初验

pressure grouting 压力灌浆

pre-stressing 预应力

pre-tensioning 先张拉

pull-off test 加强,拉拔试验

PVC water stop PVC 止水

reinforcement cover 钢筋保护层

remark 备注

resistance to freezing and thawing 抗冻融性

retractable formwork 进退式模板

ring grouting 环形灌浆

rubber water stop 橡皮止水

seal v. 密封

section area 截面面积

shear bar 剪力筋

shear strength 抗剪强度

shearkey box 键槽盒

shear-key 抗剪键

shotcrete dry-mix process 干喷混凝土生产

shotcrete wet-mix process 湿喷混凝土生产

simplify v. 简化

simultaneously adj. 同时地

single-stage grouting 一次灌浆

sliding formwork 滑模

slump n. 塌落度

soak v. 浸泡

spacer n. 水泥垫块

spacing 间距

specific weight gravity 比重

splice welding 绑条焊接

stability against sliding 抗滑稳定性

starter bar 起始筋

steel form carrier 钢模台车

steel formwork 钢模

steel waling 钢支撑

stirrup n. 箍筋

stopend n. 堵头,封堵模板

strain v. 应变

stress concentration 应力集中

III. Watching and Listening

Task One　The Gotthard Base Tunnel (I)

视频链接及文本

New Words

audacious adj. 无畏的

crisscross adj. 十字交叉的

malleable adj. 可塑的

radical adj. 激进的

shipworm n. 船蛆,凿船虫

burrow v. 挖洞穴

timber n. 木材,木料

mollusk n. 软体动物

shove v. 猛推

pulpwood n. 纸浆用木材

excrete v. 排泄,分泌

brittle adj. 易碎的

residue n. 残渣

suck v. 吸吮

grip v. 紧握;抓牢

emulate v. 效仿

rickety adj. 摇晃的

compartment n. 隔间

hail v. 欢呼

Exercises

1. *Watch the video for the first time and choose the best answers to the following questions.*

   1) In the mountains of Switzerland, engineers are undertaking a 35-mile tunnel driven

straight through the _____.

    A. Appalachian   B. Alps

    C. Rocky   D. Fuji

2) We have, very different _____ zones to get through of course.

    A. historical   B. chemical

    C. geological   D. physical

3) More than 20 of the network of tunnels _____ beneath the River Thames.

    A. parallel   B. crisscross

    C. underground   D. cross

4) By the early 19th century, Londoners had _____ all hope of having a tunnel beneath the river.

    A. achieved   B. realized

    C. thought   D. given up

5) Techniques the hard rock miners used on hard rock had failed, leading to collapse and _____.

    A. flooding   B. watering

    C. leaking   D. overflowing

2. *Watch the video again and decide whether the following statements are true or false.*

  1) Brunel observed the perfect tunnel-like structures left by ship-worms as they burrow through timber.(   )

  2) As it digs, this marine mollusk's bi-valve head and soft shell help support the creature as it moves through the wood.(   )

  3) The key is not actually to suck all the sand into the tube. (   )

  4) The mollusk then uses its front legs to grip itself to the side of the tunnel. (   )

  5) Brunel's idea started an underground revolution, some of which can still be found. (   )

3. *Watch the video for the third time and fill in the following blanks.*

Brunel dug the first-ever _____ under the River Thames using his engineering equivalent of a ship-worm. A _____ steel frame-work called a tunneling shield acts like the worm's hard head, supporting the _____ and preventing collapse. Working in individual compartments, _____ excavate just four inches before the whole tunneling shield is pushed _____ using screw jacks. A second group of men working _____ them line tunnel with bricks to _____ its collapse. Finished in 1843, Brunel's Thames Tunnel was hailed as the eighth _____ of the world with almost 50,000 visitors on opening day. Today it forms part of the London rail of _____, and it all began in this _____ chamber. Most people would have no idea that it's here. But of course, it's important to engineering and to tunnel engineering in particular, really can't be overstated.

4. *Share your opinions with your partners on the following topic for discussion.*

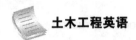
土木工程英语

Can you use a few lines to list what is your understanding about tunnel construction? Please use an example to clarify your thoughts.

Task Two　The Gotthard Base Tunnel (Ⅱ)

视频链接及文本

New Words

supersize n. 超大量

munch v.啃,咬

rotate v.轮流;使旋转

exert v.运用;发挥

disc n. 圆盘;磁盘

chip away 拆掉,削掉

conveyor n. 输送机,传送带

void n. 空隙

hydraulic adj. 水压的

shotcrete 喷射混凝土

calculate v.计算

plot v.绘图;谋划

Exercises

1. *Watch the video for the first time and choose the best answers to the following questions.*

  1) The Gotthard Base Tunnel is _____ times longer than Brunel's Thames Tunnel.

    A. 140   B. 410

    C. 14   D. 100

  2) At over 1,300 feet long, _____, or TBMs, are almost the same length as the Thames Tunnel.

    A. tunnel breaking machines   B. tunnel boring machines

    C. tube boring machines   D. tube breaking machines

  3) The TBMs need to be shipped in sections and _____ on site.

    A. loaded   B. produced

    C. assembled   D. used

  4) This is the head which is _____, and we have all in all about 60 of those discs, which are scraping the rock.

    A. swinging   B. rotating

    C. pushing   D. excavating

  5) The TBM pushes into the void using _____ legs.

    A. barometric   B. voltage

    C. steel   D. hydraulic

2. *Watch the video again and decide whether the following statements are true or false.*

  1) These machines are really very amazing. You have to imagine it's 400 meters(1,300 feet) long, the whole machine. (　)

  2) As the TBM head turns, it exerts 3.5 tons of force, chipping away the rock.(　)

  3) In the wake of the TBM, the walls are sprayed with shotcrete, a form of solid concrete. (　)

4) With this Shotcrete, lining the tunnel is safe. (　　)

5) Shotcrete is applied at the walls to help pieces of rock falling down. (　　)

3. *Watch the video for the third time and fill in the following blanks of the table.*

But how will two teams digging on _____ sides of the Swiss Alps meet in the middle? Calculating a _____ route is difficult enough _____ ground. There are trees, hills, and buildings in the _____. Using landmarks to _____ or relying on towers to get _____ sighting's helps. When _____ a course underground, even the slightest miscalculation could _____ disastrous results. Getting the TBM's meeting in the middle, it's a very, very big _____. Because, we have so _____ distances.

4. *Share your opinions with your partners on the following topic for discussion.*

1) Do you know TBMs? Can you illustrate the function of TBMs?

2) Can you describe the whole construction process of The Gotthard Base Tunnel?

IV. Talking

Task One　Classical Sentences

Directions: *In this section, some popular sentences are supplied for you to read and to memorize. Then, you are required to simulate and produce your own sentences with reference to the structure.*

General Sentences

1. My hobby is collecting stamps. Do you have a hobby?
   我的爱好是集邮。你有什么爱好?

2. I've always thought photography would be an interesting hobby.
   我一直认为摄影是一种有趣的爱好。

3. Some people like horseback riding, but I prefer golfing as a hobby.
   一些人喜欢骑马,但是我喜欢打高尔夫。

4. Do you have any special interests other than your job?
   除了工作以外,你还有什么其他特殊爱好吗?

5. Learning foreign languages is just an avocation with me.
   学外语只是我的业余爱好。

6. I find stamp collecting relaxing and it takes my mind off my work.
   我发现集邮使人放松,能让我的注意力从工作中移开。

7. On weekends I like to get my mind off my work by reading good books.
   周末我喜欢通过读好书来把我的注意力从工作上转移开。

8. My cousin is a member of a drama club. He seems to enjoy acting.
   我堂兄是戏剧俱乐部的成员。他似乎喜欢表演。

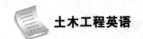

9. He plays the piano for his own enjoyment.
   他弹钢琴是为了自娱自乐。

10. I've gotten interested in Wi-Fi. I'm building my own equipment.
    我对无线网络很感兴趣。我正在安装我自己的设备。

11. He's not a professional. He plays the piano for the fun of it.
    他不是专业人士。他弹钢琴是为了好玩。

12. I've heard of unusual hobbies, but I've never heard of that one.
    我听说过一些不寻常的爱好,但我从来没听说过那一个。

13. The trouble with photography is that it's an expensive hobby.
    摄影的问题在于,它是一种昂贵的爱好。

14. That's rare set of coins. How long did it take you to collect them?
    这是一套罕见的钱币。你用了多久收集的?

15. I started a new hobby. I got tired of working in the garden.
    我开始了一个新的爱好,我厌倦了在花园里干活。

16. Baseball is my favorite sport. What's your favorite?
    棒球是我最喜欢的体育运动。你最喜欢的是什么?

17. My nephew is a baseball player. He is a catcher.
    我的外甥是一名棒球运动员。他是一名接球手。

18. When you played football, what position did you play?
    你踢足球时, 踢什么位置?

19. We played a game last night. The score was six-to-six.
    我们昨晚玩了一场比赛。比分是 6 比 6 平。

20. I went to a boxing match last night. It was a good fight.
    昨晚我去看了一场拳击比赛。比赛很精采。

21. When I was on the track team, I used to run the quarter mile.
    我在田径队时,经常跑四分之一英里。

22. I like fishing and hunting, but I don't like swimming.
    我喜欢钓鱼和打猎,但是不喜欢游泳。

23. My favorite winter sport is skiing. I belong to a ski club.
    我最喜欢的冬季运动是滑雪。我是滑雪俱乐部的成员。

24. Would you be interested in going to the horse races this afternoon?
    今天下午你有兴趣去看赛马吗?

25. What's your favorite kind of music? Do you like jazz?
    你最喜欢什么音乐? 你喜欢爵士乐吗?

26. He's a composer of serious music. I like his music a lot.
    他是一名流行音乐作曲家。我很喜欢他的音乐。

27. My brother took lessons on the trumpet for nearly ten years.
    我的哥哥练习吹喇叭将近十年了。

28. You play the piano beautifully. How many hours do you practice every day?
   你钢琴弹得很好。你每天练习多长时间？

29. I've never heard that piece before. Who wrote it?
   我从没有听过这一段。是谁写的？

30. Have you ever thought about becoming a professional musician?
   你有没有想到过要成为一名专业的音乐家？

31. Be a good sportsman and play according to the rules of the game.
   成为一个好的运动员并遵守游戏规则。

32. Our family went camping last summer. We had to buy a new tent.
   我家去年夏天去露营了。我们得买个新的帐篷。

33. This afternoon we went to the gym for a workout. We lifted weights.
   今天下午我们去体育馆健身了。我们练了举重。

34. What do you do for recreation? Do you have a hobby?
   你闲暇时做什么？你有什么爱好吗？

35. My muscles are sore from playing baseball.
   打完棒球后，我的肌肉一直酸痛。

36. I sent in a subscription to that magazine. It's put out every week.
   我订阅了那份杂志。它是周刊。

37. If you subscribe to the newspaper, it'll be delivered to your door.
   如果你订阅报纸的话，可以送到你家。

38. I didn't read the whole paper. I just glanced at the headlines.
   我没有通读全文。我只是看了看标题。

39. The first chapter of the story is in this issue of the magazine.
   这个故事的第一章刊登在本期杂志上。

40. I haven't seen the latest issue of the magazine. Is it out yet?
   我还没有看到这个杂志的最新一期。是不是还没有出版？

41. What's the total circulation of this newspaper?
   这个报纸的总发行量怎样？

42. I'm looking for the classified section. Have you seen it?
   我在找分类广告栏。你看到了吗？

43. My brother-in-law is a reporter in the *New York Times*.
   我姐夫是《纽约时报》的记者。

44. There was an article in today's paper about the election.
   今天的报纸上有选举的消息。

45. There wasn't much news in the paper today.
   今天的报纸上没有太多的消息。

46. How long have you been taking this magazine?
   你订这份杂志多久了？

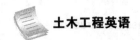

47. Did you read the article about the rescue of the two fishermen?

你读了那篇关于营救两名渔夫的文章了吗?

48. Why don't you put an advertisement in the paper to sell your car?

你为什么不在报纸上登个广告卖你的车呢?

49. I got four replies to my ad. about the bicycle for sale.

我的自行车待售广告有了四个回复。

50. My son has a newspaper route. He delivers the morning paper.

我儿子有送报的路线。他送晨报。

Specialized Sentences

1. Tunnels through hard rock are usually built by drilling and blasting.

穿过坚硬岩石的隧道通常采用钻孔和爆破。

2. A pattern of holes is drilled into the rock face by using compressed air drills operated by men on a moving carriage or "jumbo" running on temporary rails.

一种岩石钻孔方法是通过手动空压钻机在活动的支架上或能在临时铁轨上移动的巨大的钻孔机来成孔的。

3. A drill tipped with a tungsten-carbide bit can penetrate 2 to 3 meter in four to five minutes.

带有钨碳钻头的钻机能够在4至5分钟穿透2至3米岩石。

4. Each machine uses one boom to position itself inside the tunnel's coordinates by reading a laser set up inside the excavation.

每台钻机利用钻臂读取隧道内的激光坐标来定位。

5. A preset drilling pattern is then installed in the jumbo on a floppy disk, and the machine reads this information and automatically drills the holes in accordance with this information using the other booms.

接着用磁盘将预先设定的钻挖程序装入钻机,钻机就按照这个程序自动钻孔。

6. By this way the jumbos drilled holes at the exact angle and to the exact length specified by the designer, making the operation far more accurate than non-computerised techniques, as well as saving time and cutting down overbreak.

这些巨型钻机能准确地按设计要求的角度和长度钻孔施工,既比非电脑化的操作效果好,又节省了施工时间、减少了挖超。

7. When a round of holes is ready a high explosive such as dynamite is packed in and detonated.

当整块岩石周围的孔都钻成后,将高爆炸性的炸药装入并引爆。

8. A mobile rock shovel lifts the shattered rock into dump cars, which are hauled away by a locomotive.

移动式岩石装载机将岩石碎块装入可倾卸车箱里,然后由车头拉走。

9. Soft material such as sandstone, clay and chalk is cut through by automatic machines.

一些软质材料如砂石、黏土和粉岩可以通过自动机械开挖。

10. One of the largest of these machines to be used cut five 9 m tunnels through 480 m of sandstone and limestone during the building of Pakistan's Mangla Dam in 1963.

1963 年在修建巴基斯坦曼格拉大坝时,其中一台最大的设备在 480 米的砂石和石灰石岩层中开挖了 5 条直径为 9 米的隧道。

11. Machines of this kind have a hydraulic cutting head that turns slowly and scrapes out the material at the face of the tunnels.

这种机械有一个液压的钻头,慢慢地旋转把隧道前面的石料挖出来。

12. This is lifted by a mechanical shovel on to a conveyor that carries it back to dump cars behind the machine.

这由机械铲装到传动带上,然后将石料倒进在开挖机后面的倾卸车箱。

13. Mechanic alarms lift and place in position huge prefabricated sections of concrete tunnel lining.

机械臂将巨大的混凝土隧道衬里预制断面块吊起并就位。

14. Shallow tunnels are sometimes built by the "cut and cover method".

浅埋的隧道有时通过"大开挖法"施工。

15. A deep trench is excavated, the completely roofed tunnel lining is built at the bottom, and this is then covered with excavated material.

先开挖一条深沟,接着在其底部施工整条隧道,然后进行回填。

16. In underwater tunneling the working area may have to be pressurized so that the internal air pressure exceeds the pressure of water.

水下隧道施工必须进行增压以使内部空气压力超过水压。

17. After placing the tunnel lining, engineers pump cement grout, a sealing mixture, around it to make it watertight.

在隧道衬砌就位后,再泵入水泥浆密封料,以起防水作用。

18. Another method sometimes used in water bearing gravel is to sink tubes on each side of the path of the tunnel and pump in liquid nitrogen.

另外一种在水浸岩石地下开挖的方法是在隧道的两边沉管并泵入液体氮。

19. The water in the gravel is frozen solid and the tunnel can then be cut.

当砾石里的水冻结后,再开挖隧道。

20. Brunel constructed his tunnel by means of a tunneling shield—a device consisting basically of a vertical face of stout horizontal timber beams that could be removed one at a time to enable clay to be dug out, each beam then being replaced farther forward.

布鲁内尔通过隧道盾构来建造其隧道,该盾构基本上就是由水平的大梁组成的一个竖直面,大梁一次取走一根以便把土挖出来,每一根梁依次取下然后放到前面更远的地方。

21. We shall put the machine to trial (test run) after the erection work has been finished.

这台机器安装工作完成以后就将进行试车(试运转)。

22. The mechanical completion check list of the unit has been approved by both of the

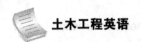
buyer's and seller's representative.

这个装置的机械竣工检验表已由买方和卖方的代表审定。

23. We should start the installation according to the instruction and operation manual.

我们应该根据说明书和操作手册来启动这个装置。

24. The systematic hydrostatic test (dry run, hot test, dynamic test, actual start-up) is scheduled for next Monday.

系统水压试验(演练、加热试验、动力试验、正式启动)定于下星期一进行。

25. We have planned to finish the adjustment of the machine before Tuesday.

我们计划在星期二以前完成机器的调试工作。

26. Before initial start-up of the installation, we must check the equipment fully.

在装置安装开始以前,我们必须仔细地检查这些设备。

27. Shall we begin the test run at once?

我们立即开始试车好吗?

28. The compressor is loaded up with the medium of air (nitrogen, process gas).

压缩机以空气(氮气、工业废气)加载运转。

29. The turbine had been running for 4 hours before carrying a full load.

涡轮机在满载前已经运转了四个小时。

30. We shall soon put the chemical installation into commissioning test run (performance test).

我们将很快地把这个化工装置进行投料试生产(性能考核)。

31. According to the schedule, the first batch process will be produced on October first this year.

根据进度表,今年 10 月 1 日将首次批量生产。

32. The machine is in good working order.

这台机器运转良好。

33. The machine is out of order; will you see to it, please?

这台机器出问题了,请你去查看一下好吗?

34. I felt the machine shake seriously.

我感到这机器振动严重。

35. The machine parts went hot.

这机器零件发热。

36. The noise of the machine is very loud.

这台机器噪声很大。

37. The machine is knocking badly.

这台机器敲击声厉害。

38. If there arises any abnormal temperature (unusual noises, vibration), it is necessary to stop the machine and investigate the cause.

如果产生不正常的温升(异常噪声、振动),必须停车查明原因。

39. You must turn off the switch when anything goes wrong with the motor.
如果电动机有什么毛病时,你必须关掉开关。

40. The rotation number of the machine is on the increase.
机器的转数在增加。

41. After a few hours running, we shall check the machine and replace the oil, if necessary.
在数小时运转后,我们将检查机器,并在必要时换油。

42. We shall select the suitable grease in accordance with the lubrication chart.
我们要根据润滑表来选用合适的油脂。

43. What is the trouble with the machine?
这机器有什么故障?

44. I think the trouble lies here.
我想故障在这里。

45. It is necessary that we should repair it at once.
我们必须立即修理它。

46. We shall give the machine another trial at 10 o'clock.
我们将在十点钟把这台机器再试一次。

47. The machine runs perfectly well, it had been operating with a continuous run of 72 hours.
这台机器运转很好,它至今已连续运转了 72 小时。

48. The result of the test runs satisfied us.
试车结果使我们很满意。

49. It is not doubtful that the test run will be successful.
毫无疑问,试车将会是成功的。

50. Fundamentally, engineering is an end-product-oriented discipline that is innovative, cost-conscious and mindful of human factors.
从根本上,工程是一个以最终产品为导向的行业,它具有创新、成本意识,同时也注意到人为因素。

Task Two　Sample Dialogue

Directions: *In this section, you are going to read several times the following sample dialogue about the relevant topic. Please pay special attention to five C's (culture, context, coherence, cohesion and critique) in the sample dialogue and get ready for a smooth communication in the coming task.*

(*Before class, two students, Huang and Valdes are talking about tunnel construction.*)

Huang: Mr. Valdes, tunneling was probably one of man's earliest exercises in the field of civil engineering. The ancient Egyptians are known to have built tunnels for transporting water and for use as tombs.

Valdes: I've heard that the Egyptians also undertook mining operations, cutting deep tunnels

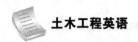

to excavate copper ores.

Huang: Yes, but today tunnels are used for road, rail and pedestrian transport, for carrying water and in mining.

Valdes: I know the first underwater tunnel was probably built in about 2160 BC by the engineers of Queen Semiramis of Babylon. The Euphrates had been diverted and the engineers dug a channel in the riverbed. In this they built a brick-lined tunnel some 900 m long, waterproofed with bitumen plaster about 2 m thick.

Huang: Yes, I read from the book that it connected the palace with a temple across the water.

Valdes: And the tunnels have often been used in warfare to penetrate enemy defenses.

Huang: Do you know that historians suggest that the walls of Jericho were almost certainly brought down by driving a tunnel beneath them and then lighting a fire to burn away the wooden props?

Valdes: Really? Tunnels, some cut through hard rock, were used extensively by the Romans in building their famous system of aqueducts. The Appian aqueduct, built in about 312 BC, ran as a tunnel for almost 25 km.

Huang: After the time of the Romans no large tunnels were built for more than a thousand years.

Valdes: Yes! It was the coming of the canal age in the seventeenth century that produced a new generation of tunnel builders.

Huang: The books says that man's first great tunnel built for transportation was part of the Canal du Midi. It was completed in 1681 and ran across France from the Bay of Biscay to the Mediterranean. At Malpus, near Beziers, a 158 m tunnel was cut to carry the canal through a rocky ridge.

Valdes: Right. It was the first tunnel built with the aid of explosives—gunpowder in hand-drilled holes.

Huang: And then throughout the eighteenth century canal tunnels were built both in Europe and America, but with the onset of the railroad age in the early nineteenth century canals fell into disuse as a means of transport.

Valdes: Yes, the construction of railroads, however, itself produced a huge increase in tunneling. One of the most remarkable and difficult tunnels was the Simplon tunnel under the Alps, completed in 1906. It runs for 20 km and connects Switzerland and Italy.

Huang: We share a lot of information on tunnel construction today. I hope we can have a talk next time.

Task Three   Simulation and Reproduction

Directions: *The class will be divided into three major groups, each of which will be assigned a topic. In each group, some students may be the teacher, while others may be students. In the process of discussion, please observe the principles of cooperation, politeness and choice of words.*

One of the groups will be chosen to demonstrate the discussion to the class.

1）Difficulties in tunnel construction.

2）The famous tunnel construction that impressed you most.

3）The importance of learning tunnel construction.

Task Four   Discussion and Debate

Directions：*The class will be divided into two groups. Please choose your stand in regard to the following controversy and support your opinions with scientific evidences. Please refer to the specialized terms and classical sentences in the previous parts of this unit.*

Read about the three basic types of tunnel construction in common use. Which one do you think can be mostly used in the tunnel construction in your city? Give your reasons.

V. After-class Exercises

1. *Match the English words in Column A with the Chinese meaning in Column B.*

A	B
1）graduation	A）钢筋保护层
2）impermeability	B）标定
3）tensile strength	C）级配
4）airtightness	D）进退式模板
5）calibration	E）湿喷混凝土生产
6）concrete release sheet	F）抗渗性
7）expansion joint	G）气密性
8）reinforcement cover	H）发料单
9）shotcrete wet-mix process	I）伸缩缝
10）retractable formwork	J）抗拉强度

2. *Fill in the following blanks with the words or phrases in the word bank. Change the forms if it's necessary.*

blast	adjust	exceed	excavation	replace
dynamite	trial	lift	approve	abnormal

1）Tunnels through hard rock are usually built by drilling and _____.

2）Each machine uses one boom to position itself inside the tunnel's coordinates by reading a laser set up inside the _____.

3）When a round of holes is ready a high explosive such as _____ is packed in and detonated.

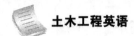
    4）Mechanic alarms _____ and place in position huge prefabricated sections of concrete tunnel lining.

    5）In underwater tunneling the working area may have to be pressurized so that the internal air pressure _____ the pressure of water.

    6）The mechanical completion check list of the unit has been _____ by both of the Buyer's and Seller's representative.

    7）If there arises any _____ temperature (unusual noises, vibration), it is necessary to stop the machine and investigate the cause.

    8）After a few hours running, we shall check the machine; and _____ the oil, if necessary.

    9）We shall give the machine another _____ at 10 o'clock.

    10）We have planned to finish the _____ of the machine before Tuesday.

3. *Translate the following sentences into English.*

    1）毫无疑问,试车将会是成功的。

    _____

    2）移动式岩石装载机将岩石碎块装入可倾卸车箱里,然后由车头拉走。

    _____

    3）这种机械有一个液压的钻头,慢慢地旋转把隧道前面的石料挖出来。

    _____

    4）浅埋的隧道有时通过"大开挖法"施工。

    _____

    5）当砾石里的水冻结后,再开挖隧道。

    _____

4. *Please write an essay of about 120 words on the topic: The Difficulties in Tunnel Construction. Some specific examples will be highly appreciated and watch out the spelling of some specialized terms you have learned in this unit.*

    _____

    _____

    _____

    _____

    _____

VI. Additional Reading

Brief Introduction on Tunnel construction

Cut-and-cover

[A] Cut-and-cover is a simple method of construction for shallow tunnels where a trench(沟渠) is excavated and roofed over with an overhead support system strong enough to carry the load of what is to be built above the tunnel. Two basic forms of cut-and-cover tunneling are available:

Bottom-up method: A trench is excavated, with ground support as necessary, and the tunnel is constructed in it. The tunnel may be of in situ concrete, precast concrete, precast arches, or corrugated(起波纹) steel arches; in early days brickwork was used. The trench is then carefully back-filled and the surface is reinstated.

Top-down method: Side support walls and capping beams are constructed from ground level by such methods as slurry(泥浆) walling or contiguous bored piling. Then a shallow excavation allows making the tunnel roof of precast beams or in situ concrete. The surface is then reinstated except for access openings. This allows early reinstatement of roadways, services and other surface features. Excavation then takes place under the permanent tunnel roof, and the base slab(厚板) is constructed.

[B] Shallow tunnels are often of the cut-and-cover type (if under water, of the immersed-tube type), while deep tunnels are excavated, often using a tunnelling shield. For intermediate levels, both methods are possible.

[C] Large cut-and-cover boxes are often used for underground metro stations, such as Canary Wharf tube station in London. This construction form generally has two levels, which allows economical arrangements for ticket hall, station platforms, passenger access and emergency egress, ventilation and smoke control, staff rooms, and equipment rooms. The interior of Canary Wharf station has been likened to an underground cathedral, owing to the sheer size of the excavation. This contrasts with many traditional stations on London Underground, where bored tunnels were used for stations and passenger access. Nevertheless, the original parts of the London Underground network, the Metropolitan and District Railways, were constructed using cut-and-cover. These lines pre-dated electric traction and the proximity to the surface was useful to ventilate(通风) the inevitable smoke and steam.

[D] A major disadvantage of cut-and-cover is the widespread disruption generated at the surface level during construction. This, and the availability of electric traction, brought about London Underground's switch to bored tunnels at a deeper level towards the end of the 19th century.

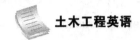

Boring machines

[E] Tunnel boring machines (TBMs) and associated back-up systems are used to highly automate the entire tunnelling process, reducing tunnelling costs. In certain predominantly urban applications, tunnel boring is viewed as quick and cost effective alternative to laying surface rails and roads. Expensive compulsory purchase of buildings and land, with potentially lengthy planning inquiries, is eliminated. Disadvantages of TBMs arise from their usually large size—the difficulty of transporting the large TBM to the site of tunnel construction, or (alternatively) the high cost of assembling the TBM on-site, often within the confines of the tunnel being constructed.

[F] There are a variety of TBM designs that can operate in a variety of conditions, from hard rock to soft water-bearing ground. Some types of TBMs, the bentonite (膨润土) slurry and earth-pressure balance machines, have pressurized compartments at the front end, allowing them to be used in difficult conditions below the water table. This pressurizes the ground ahead of the TBM cutter head to balance the water pressure. The operators work in normal air pressure behind the pressurized compartment, but may occasionally have to enter that compartment to renew or repair the cutters. This requires special precautions, such as local ground treatment or halting the TBM at a position free from water. Despite these difficulties, TBMs are now preferred over the older method of tunnelling in compressed air, with an air lock/decompression chamber some way back from the TBM, which required operators to work in high pressure and go through decompression procedures at the end of their shifts, much like deep-sea divers.

[G] In February 2010, Aker Wirth delivered a TBM to Switzerland, for the expansion of the Linth-Limmern Power Stations located south of Linthal in the canton of Glarus. The borehole has a diameter of 8.03 meters (26.3 ft.). The four TBMs used for excavating the 57-kilometre Gotthard Base Tunnel, in Switzerland, had a diameter (直径) of about 9 meters (30 ft.). A larger TBM was built to bore the Green Heart Tunnel (Dutch: Tunnel Groene Hart) as part of the HSL-Zuid in the Netherlands, with a diameter of 14.87 meters (48.8 ft.). This in turn was superseded by the Madrid M30 ringroad, Spain, and the Chongming tunnels in Shanghai, China. All of these machines were built at least partly by Herrenknecht. As of August 2013, the world's largest TBM is "Big Bertha", a 57.5-foot (17.5 meters) diameter machine built by Hitachi Zosen Corporation, which is digging the Alaskan Way Viaduct replacement tunnel in Seattle, Washington (US).

Clay-kicking

[H] Clay-kicking is a specialized method developed in the United Kingdom of digging tunnels in strong clay-based soil structures. Unlike previous manual methods of using mattocks (鹤嘴锄) which relied on the soil structure to be hard, clay-kicking was relatively silent and hence did not harm soft clay-based structures. The clay-kicker lies on a plank at a 45-degree angle away from the working face and inserts a tool with a cup-like rounded end with the feet. Turning the tool manually, the kicker extracts a section of soil, which is then placed on the waste extract.

Used in Victorian civil engineering, the method found favor in the renewal of Britain's ancient sewerage(排水系统)systems, by not having to remove all property or infrastructure to create a small tunnel system. During the First World War, the system was used by Royal Engineer tunnelling companies to put mines beneath the German Empire lines. The method was virtually silent and so not susceptible(敏感)to listening methods of detection.

Shafts

[I]A temporary access shaft is sometimes necessary during the excavation of a tunnel. They are usually circular and go straight down until they reach the level at which the tunnel is going to be built. A shaft normally has concrete walls and is usually built to be permanent. Once the access shafts are complete, TBMs are lowered to the bottom and excavation can start. Shafts are the main entrance in and out of the tunnel until the project is completed. If a tunnel is going to be long, multiple shafts at various locations may be bored so that entrance to the tunnel is closer to the unexcavated area.

Once construction is complete, construction access shafts are often used as ventilation shafts, and may also be used as emergency exits.

Sprayed concrete techniques

[J]The New Austrian Tunneling Method (NATM) was developed in the 1960s and is the best known of a number of engineering practices that use calculated and empirical measurements to provide safe support to the tunnel lining. The main idea of this method is to use the geological stress of the surrounding rock mass to stabilize the tunnel, by allowing a measured relaxation and stress reassignment into the surrounding rock to prevent full loads becoming imposed on the supports. Based on geotechnical measurements,an optimal(最优化)cross section is computed. The excavation is protected by a layer of sprayed concrete, commonly referred to as shotcrete. Other support measures can include steel arches, rock-bolts and mesh. Technological developments in sprayed concrete technology have resulted in steel and polypropylene(聚丙烯) fibers being added to the concrete mix to improve lining strength. This creates a natural load-bearing ring, which minimizes the rock's deformation.

[K]By special monitoring the NATM method is flexible, even at surprising changes of the geomechanical rock consistency during the tunneling work. The measured rock properties lead to appropriate tools for tunnel strengthening. In the last decades also soft ground excavations up to 10 kilometres became usual.

1. *Read the passage with ten sentences attached to it. Each statement contains information given in one of the paragraphs. Identify the paragraph from which the information is derived. You may choose a paragraph more than once. Each paragraph is marked with a letter.*

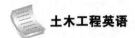

1) This and the availability of electric traction, brought about London Underground's switch to bored tunnels at a deeper level towards the end of the 19th century.

2) This pressurizes the ground ahead of the TBM cutter head to balance the water pressure.

3) Shallow tunnels are often of the cut-and-cover type (if under water, of the immersed-tube type), while deep tunnels are excavated, often using a tunnelling shield.

4) The tunnel may be of in situ concrete, precast concrete, precast arches, or corrugated steel arches; in early days brickwork was used.

5) The New Austrian Tunneling Method (NATM) was developed in the 1960s and is the best known of a number of engineering practices

6) Unlike previous manual methods of using mattocks which relied on the soil structure to be hard, clay-kicking was relatively silent and hence did not harm soft clay-based structures.

7) Expensive compulsory purchase of buildings and land, with potentially lengthy planning inquiries, is eliminated.

8) A larger TBM was built to bore the Green Heart Tunnel (Dutch: Tunnel Groene Hart) as part of the HSL-Zuid in the Netherlands, with a diameter of 14.87 meters (48.8 ft.).

9) The interior of Canary Wharf station has been likened to an underground cathedral, owing to the sheer size of the excavation.

10) By special monitoring the NATM method is flexible, even at surprising changes of the geotechnical rock consistency during the tunneling work.

1) _____ 2) _____ 3) _____ 4) _____ 5) _____
6) _____ 7) _____ 8) _____ 9) _____ 10) _____

2. *In this part, the students are required to make an oral presentation on either of the following topics.*

1) Illustrate the function of the Tunnel boring machines.

2) What impressed you deeply from the above information of tunnel construction?

习题答案

Unit Ten   The Forbidden City

I. Pre-class Activity

Directions: *Please read the general introduction about **The Forbidden City** and tell something more about the great architecture to your classmates.*

The Forbidden City

The Forbidden City is a palace complex in central Beijing, China. The former Chinese imperial palace from the Ming dynasty to the end of the Qing dynasty (the years from 1420 to  1912), it now houses the Palace Museum. The Forbidden City served as the home of emperors and their households as well as the ceremonial and political center of Chinese government for almost 500 years.

Constructed from 1406 to 1420, the complex consists of 980 buildings and covers 72 hectares (over 180 acres). The palace exemplifies traditional Chinese palatial architecture, and has influenced cultural and architectural developments in East Asia and elsewhere. The Forbidden City was declared a World Heritage Site in 1987, and is listed by UNESCO as the largest collection of preserved ancient wooden structures in the world.

Since 1925, the Forbidden City has been under the charge of the Palace Museum, whose extensive collection of artwork and artifacts were built upon the imperial collections of the Ming and Qing dynasties. Part of the museum's former collection is now in the National Palace Museum in Taipei. Both museums descend from the same institution, but were split after the Chinese Civil War. Since 2012, the Forbidden City has seen an average of 15 million visitors annually, and received more than 16 million visitors in 2016 and 2017.

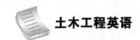土木工程英语

II. Specialized Terms

Directions：*Please memorize the following specialized terms before the class so that you will be able to better cope with the coming tasks.*

a round of excavation 一个开挖循环
adit n. 支洞,通路
advance n. 进尺
average quantity used in unit 平均单耗
backfill v.回填
blasting result 爆破结果
blend v.混合
blockness n. 块状
bottom drift 底部掏槽
breaking hole 塌落孔
center drift 中心掏槽
coefficient of non-uniformity 开挖不均匀
　系数
compact v.碾压
concrete excavation 混凝土开挖
controlled perimeter blasting 周边控制爆破
controlled sprinkling 控制洒水
cut the slope 削坡
data of explosive filled 装药参数
data of holes drilled 钻孔参数
delayed blasting 延时爆破
dewatering n. 排水
distance between holes 孔距
distance between rows 排距
drilling n. 钻孔
drilling and blasting 钻孔爆破
driving (progress) rate 进尺率
dump v.卸料
easer n. 掏槽孔
erecting supports for the roof and side wall
　对顶拱及边墙进行支护
excavation and support 开挖与支护

excavation pit 开挖基坑
expansion shell rock bolt 胀壳式张拉锚杆
exploratory hole 探孔
explosive quantity in a sound 单响药量
explosive quantity 药量
fault excavation 断层开挖
feature rock anchor 随机锚索
feature rock bolt 随机张拉锚杆
feature rock dowel 随机砂浆锚杆
feature rock reinforcement 随机加固
foundation excavation 基础开挖
full face excavation 全断面开挖
handling misfire 哑炮处理
heading and bench excavation 导洞与台阶
　开挖
intensity of excavation 开挖强度
lattice girder 钢桁架,格构大梁
light charge 少量装药
loading (charging)装药
loading and hauling muck (mucking) 装拉
　渣土
local excavation 局部开挖
loosening blasting 松动爆破
mass (bulk) excavation 大面积开挖
overexcavate (overbreak) v.超挖
pattern n. (布孔)方式
pattern cement grouted rock dowel 水泥浆
　锚杆系统
pattern resin grouted rock dowel 树脂浆锚
　杆系统
pattern rock dowel 砂浆锚杆系统
pattern rock reinforcement 加固系统

perimeter hole 周边孔

permanent rock support 永久支护

pilot tunnel 导洞

post shearing blasting 微差爆破

post-tensioned cement grouted tendon rock anchor 后张拉水泥浆锚索

protective layer excavation 保护层开挖

pump sump 水泵坑

quantity of holes 孔数

rebound material 回弹料

removing dust 除尘

removing ground water 清除地面积水

return rock bolt 回程锚杆

return rock reinforcement 回岩加固

rock excavation 岩石开挖

rock support 岩石支护

shooting the explosive(blasting) 放炮

shotcrete anchorage 喷锚

shotcrete with wire mesh 挂网喷射混凝圭

side drift 边部掏槽

slope excavation 边坡开挖

smooth blasting 光面爆破

soft ground excavation 软基开挖

soil excavation 土方开挖

steel fabric shotcrete 钢纤维喷射混凝土

steel rib 钢拱架

structural excavation 结构开挖

temporary rock support 临时支护

tensioned resin grouted rock bolt 树脂张拉锚杆

tooth excavation 齿槽开挖

top drift 顶部掏槽

top heading excavation 上导洞开挖

tunnel and underground excavation 洞挖及地下开挖

underbreak v.欠挖

unclassified material excavation 不分类料开挖

underbreak treatment 欠挖处理

ventilation 通风

water pressure test 水压试验

water stop 止水

water-cement ratio 水灰比

water-tightness 闭水性

winter protection 冬季保温

wood formwork 木模

III. Watching and Listening

Task One   Secrets of China's Forbidden City（I）

视频链接及文本

New Words

edifice n. 大建筑物

vast adj. 巨大的

golden floor tiles 金箔包裹的砖

gold leaf 金叶

garrison adj. 守备部队

grand canal 大运河

water diversion project 引水项目

reign vi. 当政，统治

tank n. 油/水箱，罐，槽

barge n. 驳船

slice n. 薄片；部分

vessel n. 容器，器皿

conscript vt. 征兵

rudimentary a.基本的,初步的

float vi. 浮动，飘动

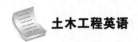

Exercises

1. *Watch the video for the first time and choose the best answers to the following questions.*

1) Over 100,000 Nanmu logs from 1,800 kilometers to the _____.

    A. south east   B. south west

    C. north west   D. south

2) Gold leaves are brought from _____.

    A. Nanjing   B. Beijing

    C. the south   D. the east

3) Rebuilding a canal that runs over 1,700 kilometers, it then links China's lifeblood rivers the _____ and Yellow.

    A. Amazon   B. Changjiang

    C. Yangtze   D. Zhujiang

4) The river water pours in to _____ the ships in the lock so they can then proceed.

    A. raise   B. decline

    C. stop   D. water

5. The grand canal allowed engineers to _____ the Nanmu north to Beijing.

    A. move   B. carry

    C. ship   D. float

2. *Watch the video again and decide whether the following statements are true or false.*

1) To supply this vast building site, huge quantities of specialized materials were exported from across the whole empire. (　)

2) Golden floor tiles are brought from 1,000 kilometers south. (　)

3) The grand canal is not longer and older than the Panama Canal and the Suez Canal. (　)

4) But the flat terrain blocked the canal, so engineers devised a vast system of locks. (　)

5) To raise the ships, the canal is fed from one river, and three huge lakes were dug as storage tanks. (　)

3. *Watch the video for the third time and fill in the following blanks.*

But to get everything to Beijing, Yongle's _____ had to dream up another engineering marvel. China's _____. The grand canal is both longer and _____ than the Panama Canal and the Suez Canal. They really _____ a very large problem of canal engineering at the time. Rebuilding a canal that _____ over 1,700 kilometers, from commercial Hangzhou in the south, it then links China's _____ rivers the Yangtze and Yellow, right up to Beijing. The water flow is _____ from Nanwang. Here we are at the center of the Nanwang water _____ project of the grand canal system. Here at Nanwang, a whole river was diverted to _____ the canal. Today, it is dry. But this is what it would have looked like in Yongle's _____.

4. *Share your opinions with your partners on the following topic for discussion.*

1) Do you like this video clip of the Forbidden City? Why do you enjoy it? Please summarize the features of Chinese traditional buildings.

2) Can you use a few lines to list what's your understanding about the difference between Chinese buildings and western buildings? Please use an example to clarify your thoughts.

Task Two   Secrets of China's Forbidden City (II)

视频链接及文本

New Words

carpenter n. 木工,木匠

replicate vt. 复制,复写

calibrate vt. 校准

timber-frame core 木框架结构

column n. 圆柱

mash n. 网状结构

elaborate adj. 精美的

versatile adj. 多用途的

buddhist temples 佛教的寺庙

palatial architecture 宫殿建筑

zenith n. 顶点,顶峰

assemble v.装配

glue n. 胶

ingenuity n. 心灵手巧

flexible adj. 灵活的

tremendous adj. 极大的

eave n. 屋檐

free-standing 活动的

vulnerable adj. 脆弱的

snap vt. & vi. 折断

fracture vi. 断裂

topple vi. 倾倒,倒塌

catastrophic adj. 灾难的

Exercises

1. *Watch the video for the first time and choose the best answers to the following questions.*

1) To be as accurate as possible, a unique scale model is being made using traditional _____ tools and techniques.

    A. wooden   B. carpentry

    C. stone   D. golden

2) The scale is one to _____.

    A. one   B. three

    C. five   D. seven

3) A fifth of the original size engineers have built it on a _____ table to simulate earthquake forces.

    A. shake   B. stone

    C. stable   D. quick

4) These beams are there to support these huge red _____ that bear the structural load of that massive roof.

    A. mash   B. eaves

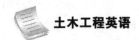

C. decorated dragons and phoenixes   D. columns

5) A _____ is a complex bracket that supports the huge roof.

A. beam   B. carpenter

C. dougong   D. structure

2. *Watch the video again and decide whether the following statements are true or false.*

1) These beams and columns are the structural core of all the Forbidden City's buildings. ( )

2) Dougong have been used throughout Western architecture for literally thousands of years. ( )

3) Dougong dated back to at least 2,000 years. ( )

4) The complicated dougong brackets extend to the interior to support the eaves as well as the roof. ( )

5) The columns are founded in the earth and are stable-standing. ( )

3. *Watch the video for the third time and fill in the following blanks.*

Carpenter Richard Weiborg finds it's easy to take one apart, but hard to _____.

"Now, after taking a deep _____, we'll try and put this back together again. And amazingly, there are no nails or _____, nothing holding it together, other than _____ ingenuity. You can see this isn't _____ for somebody who has done it a few times. It's complicated. One piece goes in the _____ here. There you have it, a bracket _____. So you can see, it's a little bit _____. It's very strong, pressing down it can take a tremendous amount of _____. It ties together with other parts of the building and is a beautiful _____ feature of Chinese architecture."

4. *Share your opinions with your partners on the following topic for discussion.*

1) Do you know dougong? How do you know this special structure?

2) Can you use a paragraph to illustrate the structure of Dougong, this beautiful and creative type of Chinese architecture?

IV. Talking

Task One   Classical Sentences

Directions: *In this section, some popular sentences are supplied for you to read and to memorize. Then, you are required to simulate and produce your own sentences with reference to the structure.*

General Sentences

1. What channel did you watch on television last night?

昨天晚上你看的哪个频道的电视?

2. I don't get a good picture on my TV set. There's probably something wrong.

我的电视机上画面不清晰。电视可能出毛病了。

3. You get good reception on your radio.

你的收音机接收效果很好。

4. Please turn the radio up. It's too low.

请把收音机开大点声。声音太小了。

5. What's on following the news and weather?

新闻和天气预报后是什么节目？

6. Do you have a TV guide?

你有电视节目指南吗？

7. You ought to have Bill look at your TV. Maybe he could fix it.

你应该让比尔检查下你的电视。他可能会修理。

8. We met one of the engineers over at the television station.

我们在电视台遇见了一个在那里工作的工程师。

9. Where can I plug in the TV? Is this outlet all right?

电视插头该插在哪里？这个插头可以用吗？

10. I couldn't hear the program because there was too much static.

因为干扰太大，我听不清节目了。

11. Your car radio works very well. What kind is it?

你的车载收音机性能很好。它是什么类型的？

12. The next time I buy a TV set, I'm going to buy a portable model.

下一次我买电视机时，我打算买一部手提式的。

13. I wonder if this is a local broadcast.

我想知道这是不是本地广播。

14. You'd get better TV reception if you had an outside antenna.

如果你有外部天线的话，你电视机的接收效果将会很好。

15. Most amateur radio operators build their own equipment.

大多数业余收音机爱好者都自己装收音设备。

16. Station Voice of WIT is off the air now. They signed off two hours ago.

武工之声电台已经停止广播了。它们在两小时之前就结束了。

17. Who is the author of this novel?

这个小说的作者是谁？

18. I've never read a more stirring story.

我从没有读过这么感人的故事。

19. Who would you name as the greatest poet of our time?

你认为谁是我们这个时代最伟大的诗人？

20. This poetry is realistic. I don't care for it very much.

这篇诗集是写实的。我不太喜欢。

21. Many great writers were not appreciated fully while they were alive.

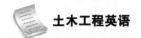

许多大作家在世时并没有得到人们的充分赏识。

22. This is a poem about frontier life in the United States.

这是一首描述美国边境生活的诗。

23. This writer uses vivid descriptions in his writings.

这位作者作品中的描述非常生动。

24. How much do you know about the works of Henry Wadsworth Longfellow?

关于亨利·沃兹沃思·朗费罗的著作你了解多少?

25. As the saying goes, where there is a will, there is a way.

俗话说,有志者事竟成。

26. It is well-known to all that all roads lead to Rome.

众所周知,条条大路通罗马。

27. Whatever is worth doing is worth doing well.

任何值得做的,就把它做好。

28. The hardest thing to learn is to be a good loser.

最难学的是做一个输得起的人。

29. Happiness is a way station between too much and too little.

幸福是位于太多和太少之间的一个小站。

30. The hard part isn't making the decision. It's living with it.

做出决定并不困难。困难的是接受决定。

31. You may be out of my sight, but never out of my mind.

你也许已走出我的视线,但从未走出我的思念。

32. Love is not a maybe thing. You know when you love someone.

爱不是什么可能、大概、也许。一旦爱上了,自己是十分清楚的。

33. In the end, it's not the years in your life that count. It's the life in your years.

到头来,你活了多少岁不算什么,重要的是,你是如何度过这些岁月的。

34. When the whole world is about to rain, let's make it clear in our heart together.

当全世界约好一起下雨,让我们约好一起在心里放晴。

35. It's better to be alone than to be with someone you're not happy to be with.

宁愿一个人待着,也不要跟不合拍的人待一块。

36. One needs three things to be truly happy living in the world: something to do, someone to love, and something to hope for.

要得到真正的快乐,我们只需拥有三样东西:有想做的事、有值得爱的人、有美丽的梦。

37. No matter how badly your heart has been broken, the world doesn't stop for your grief. The sun comes right back up the next day.

不管你有多痛苦,这个世界都不会为你停止转动,太阳依旧照样升起。

38. Accept what was and what is, and you'll have more positive energy to pursue what will be.

只有接受过去和现在,才会有力量去追寻自己的未来。

39. Until you make peace with who you are, you'll never be content with what you have.
除非你能和真实的自己和平相处,否则你永远不会对已拥有的东西感到满足。

40. If you wish to succeed, you should use persistence as your good friend, experience as your reference, prudence as your brother and hope as your sentry.
如果你希望成功,当以恒心为良友、以经验为参谋、以谨慎为兄弟、以希望为哨兵。

41. Great minds have purpose, others have wishes.
杰出的人有着目标,普通人只有愿望。

42. Being single is better than being in an unfaithful relationship.
比起谈着充满欺骗的恋爱,单身反而更好。

43. If you find a path with no obstacles, it probably doesn't lead anywhere.
太容易的路,可能根本就不会带你去任何地方。

44. Getting out of bed in winter is one of life's hardest mission.
冬天起床是人生最艰难的事情之一。

45. The future is scary but you can't just run to the past cause it's familiar.
未来会让人心生畏惧,但是我们却不能因为习惯了过去,就逃回过去。

46. Success is the ability to go from one failure to another with no loss of enthusiasm.
要想成功,你要能够跨过一个又一个失败,但仍然保持激情。

47. Not everything that is faced can be changed, but nothing can be changed until it is faced.
并不是你面对了,事情都能改变;但是,如果你不肯面对,那什么也变不了。

48. If they throw stones at you, don't throw back. Use them to build your own foundation instead.
如果别人朝你扔石头,就不要扔回去了,留着做你建高楼的基石吧!

49. If your happiness depends on what somebody else does, I guess you do have a problem.
如果你的快乐与否取决于别人做了什么,我想,你真的有点问题。

50. Today, give a stranger one of your smiles. It might be the only sunshine he sees all day.
今天,给一个陌生人送上你的微笑吧,很可能,这是他一天中见到的唯一的阳光。

Specialized Sentences

1. The importance of scheduling in insuring the effective coordination of work and the attainment of project deadlines is indisputable.
排期对于确保工作的有效协调和项目按时完成的重要性是不容置疑的。

2. For large projects with many parties involved, the use of formal schedules is indispensable.
对于涉及许多参与方的大的项目来说,使用进度计划表必不可少。

3. The network model for representing project activities has been provided as an important conceptual and computational framework for planning and scheduling.
项目活动网络模型已经为计划和编制进度提供了一个概念性的计算框架。

4. Networks not only communicate the basic precedence relationships between activities, they

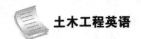

also form the basis for most scheduling computations.

网络图不仅传达两个活动之间的前后顺序关系,而且是计算大多数计划进度的基础。

5. As a practical matter, most project scheduling is performed with the critical path scheduling method, supplemented by heuristic procedures used in project crash analysis or resource constrained scheduling.

实际应用当中,大多数项目先用关键线路法编制基准进度计划,在此基础上再进行时间压缩或凭直觉和经验资源进行分析。

6. Many commercial software programs are available to perform these tasks.

有许多商业软件程序可以完成这些试验。

7. Probabilistic scheduling or the use of optimization software to perform time/cost trade-offs is more infrequently applied, but there are software programs available to perform these tasks if desired.

利用概率进度计划或最优软件来进行时间成本计算很少用到,但是需要的话也有软件程序实现这些任务。

8. Information representation is an area which can result in substantial improvements.

信息呈现是一个可以进行大幅改善的领域。

9. While the network model of project activities is an extremely useful device to represent a project, many aspects of project plans and activity inter-relationships cannot or have not been represented in network models.

尽管项目工作的网络模型是表达项目的极为有用的工具,但在项目计划和工作之间的相互关系的很多方面不能或还没有在网络模型中表示。

10. The similarity of processes among different activities is usually unrecorded in the formal project representation.

通常,在正式的项目表示中,不同活动之间流程的类似之处是没有记录的。

11. Updating a project network in response to new information about a process such as concrete pours can be tedious.

不断更新一个项目的网络图以反映某一个特定工作(比如混凝土浇筑)的新信息可能非常繁琐。

12. What is needed is a much more flexible and complete representation of project information.

所需要的是一个更加灵活和完整的项目信息。

13. Actual projects involve a complex inter-relationship between time and cost.

实际项目中时间和成本之间的关系十分复杂。

14. As projects proceed, delays influence costs and budgetary problems may in turn require adjustments to activity schedules.

随着项目的进展,进度拖延会影响到成本;反过来预算上的问题也要对进度做出某些调整。

15. Additional resources applied to a project activity might result in a shorter duration but higher costs.

在项目计划中,若为项目某个工作增加资源就会缩短时间,但同时会增加成本。

16. Unanticipated events might result in increases in both time and cost to complete an activity.

意外事件可能导致竣工的时间和成本的增加。

17. Excavation problems may easily lead to much lower than anticipated productivity on activities requiring digging.

挖掘中出现的问题可能容易导致挖掘的生产率大大低于预期。

18. The difficulty of integrating schedule and cost information stems primarily from the level of detail required for effective integration.

整合进度和成本的困难决定于有效综合所需要信息的详细程度。

19. Usually, a single project activity will involve numerous cost account categories.

通常,一个单一的项目会涉及许多成本账目。

20. An activity for the preparation of a foundation would involve laborers, cement workers, concrete forms, concrete, reinforcement, transportation of materials and other resources.

一项为基础设施做准备的活动会包括劳动力、水泥工人、混凝土模板、混凝土浇筑、加固、材料的运输和其他资源。

21. Even a more disaggregated activity definition such as erection of foundation forms would involve numerous resources such as forms, nails, carpenters, laborers, and material transportation.

即使是像安装基础模板这样的零散工作也会涉及许多资源,比如模板、钉子、木工、劳动力和材料运输。

22. Numerous activities might involve expenses associated with particular cost accounts.

许多活动可能涉及一些与特殊成本账目相关的花费。

23. A particular material such as standard piping might be used in numerous different schedule activities.

一种特殊材料(比如标准管材)会在许多不同的工作排期中使用。

24. To integrate cost and schedule information, the disaggregated charges for specific activities and specific cost accounts must be the basis of analysis.

为了综合成本和进度信息,必须把针对具体工作和具体成本而产生的分散计算的数额当作分析的基础。

25. In the absence of a work element accounting system, costs associated with particular activities are usually estimated by summing expenses in all cost accounts directly related to an activity plus a proportion of expenses in cost accounts used jointly by two or more activities.

在没有工作要素会计系统的情况下,首先汇总成本报告中与一项工作直接有关的所有费用,再加上按比例分配成本表中两个或多个工作的共同费用中与具体工作有关

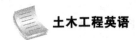

的成本。

26. The basis of cost allocation would typically be the level of effort or resource required by the different activities.

成本分配的基础原则是考虑不同活动所耗费的投入或资源的大小。

27. Costs associated with supervision might be allocated to different concreting activities on the basis of the amount of work (measured in cubic yards of concrete) in the different activities.

在不同的混凝土作业中,与监督相关的成本会根据工作量(以混凝土方量计算)分配至不同的具体活动上。

28. With these allocations, cost estimates for particular work activities can be obtained.

通过这些分配,可以获得详细的工作活动的成本估算。

29. Quality control and safety represent increasingly important concerns for project managers.

质量控制和安全是项目经理越来越关注的重点。

30. Defects or failures in constructed facilities can result in very large costs.

施工设施中的缺陷或失败可以造成非常大的成本。

31. Even with minor defects, re-construction may be required and facility operations impaired.

有时即使是较小的缺陷,也需要重新施工,并会导致设施的损坏。

32. Increased costs and delays are the result.

成本的增加和工期的拖延就是后果。

33. In the worst case, failures may cause personal injuries or fatalities.

在最坏的情况下,失误会造成人身伤害或死亡。

34. Accidents during the construction process can similarly result in personal injuries and large costs.

施工过程中的意外事故可以导致人身伤害和高昂的支出。

35. Indirect costs of insurance, inspection and regulation are increasing rapidly due to these increased direct costs.

由于这些直接成本的增加,保险、检查和管控等非直接成本也随之增加。

36. Good project managers try to ensure that the job is done right the first time and that no major accidents occur on the project.

称职的项目经理努力保证一次施工便符合要求,同时项目不出现大的事故。

37. As with cost control the most important decisions regarding the quality of a completed facility are made during the design and planning stages rather than during construction.

与成本控制一样,有关完工设施质量的重要决策是在设计和规划阶段而不是在建设期间做出的。

38. It is during these preliminary stages that component configurations, material specifications and functional performance are decided.

正是在这些初步阶段,决定了部件的配置、材料规格和功能性能。

39. Quality control during construction consists largely of insuring conformance to these original designs and planning decisions.

施工过程中的质量控制主要包括确保这些原始设计和规划决策保持一致。

40. While conformance to existing design decisions is the primary focus of quality control, there are exceptions to this rule.

虽然保持与现有设计决策一致是质量控制的主要焦点,但是这个规则也有例外。

41. Unforeseen circumstances, incorrect design decisions or changes desired by an owner in the facility function may require re-evaluation of design decisions during the course of construction.

不可预见的情况、不正确的设计决策、项目业主变更等情况,都可能要求施工者在施工过程中重新评估设计决策。

42. While these changes may be motivated by the concern for quality, they present occasions for re-design with all the attendant objectives and constraints.

虽然这些变化可能是出于对质量的关注,但它们提供了重新设计的机会,并附带了所有随之而来的目标和约束。

43. Some designs rely upon informed and appropriate decision making during the construction process itself.

一些设计依赖于施工过程中的明智、适当的决策。

44. Some tunneling methods make decisions about the amount of the tunneling.

隧道施工方法决定隧道的工程量。

45. Since such decisions are based on better information concerning actual site conditions, the facility design may be more cost effective as a result.

由于此类决策基于与实际场地条件有关的更完善信息,因此现场设计可能更具成本效益。

46. Safety during the construction project is also influenced in large part by decisions made during the planning and design process.

建设项目的安全在很大程度上也受到规划和设计过程中所作决策的影响。

47. Some designs or construction plans are inherently difficult and dangerous to implement, whereas other, comparable plans may considerably reduce the possibility of accidents.

由于固有的因素,一些设计或施工计划难以实施且很危险,而一些类似的计划可能大大减少事故的可能性。

48. Clear separation of traffic from construction zones during roadway rehabilitation can greatly reduce the possibility of accidental collisions.

在道路修复过程中,将交通与施工区域明确分开,可以大大减少意外碰撞的可能性。

49. Beyond these design decisions, safety largely depends upon education, vigilance and cooperation during the construction process.

除了这些设计决策之外,安全在很大程度上取决于施工过程中的不断培训、重视与配合。

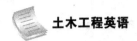

50. Workers should be constantly alert to the possibilities of accidents and avoid taken unnecessary risks.

工人应时刻警惕可能发生的事故, 避免承担不必要的风险。

51. Quality control should be a primary objective for all the members of a project team.

质量控制应该是项目团队所有成员的首要目标。

Task Two Sample Dialogue

Directions: *In this section, you are going to read several times the following sample dialogue about the relevant topic. Please pay special attention to five C's (culture, context, coherence, cohesion and critique) in the sample dialogue and get ready for a smooth communication in the coming task.*

On-site Training—Maintenance and Trouble Shooting

(Mr. Sunaga is showing the operators how to maintain the new machine and shoot troubles.)

Sunaga: Today we'll see how to do the trouble shooting and maintain the machine. All of our machines have passed the strict inspections and tests before sale and can be used safely. However, the performance, safety, working efficiency and service life greatly depend on your daily handling and maintenance. Have you got the Operation Manuals with you?

Operator: Yes.

Sunaga: Fine. We'll refer to the Manual as we go along. As I said just now, regular maintenance is very important in ensuring that the machine is always operated in the best conditions. Let's see what we have to do to make the machine work well. Again, change the hydraulic oil if it is found to be contaminated or moisture is mixed in it. This is the first thing you must keep in mind because hydraulic oil plays a very important role in the machine's functioning. As you know, this is a hydraulic jumbo. You have to replace the oil filters and the air cleaners atregular intervals.

Operator: How often do we replace them?

Sunaga: For the oil filters, 50 working hours for the first time on a new machine and after that, every 300 hours. For the air cleaners, every 400 hours. It's better to clean them during the interval. Now let's move on to the drive unit area. You grease the center bearing every 40 working hours. Next is how to do the trouble shooting. Sometimes problems occur during the operation. First you try to find where the trouble lies. Suppose the grade ability is poor. Where do we have to check?

Operator: We have to see if the parking brake and the foot brake are released completely.

Sunaga: If yes, then where do we go next?

Operator: I am not sure, perhaps...

Sunaga: You can find the right answer in the trouble shooting flowchart on page 81 of the Manual. That is, we go on to check whether there are any oil leakages, whether the hydraulic oil level is too low or contaminated or the suction filter is clogged. So, always refer to the Chart for help if you're not sure of how to locate the trouble.

Operator: How do we know it's time to overhaul the machine?

Sunaga: Normally, only the units, like the engine, the two drifters and the compressors need to be overhauled. The other parts need to be checked and maintained regularly only. You can find what need to be overhauled, and how many hours they work before they should be overhauled, in the Manual. To control lever overhaul the units, you need to work together with your mechanical engineer in this field or ask us for help, for some of the procedures are quite complicated and special tools are needed. Any questions?

Operator: We don't have any for the moment.

Sunaga: OK. Tomorrow, we will practice drilling the rock with the machine.

Operator: We're eager to try for ourselves.

Task Three   Simulation and Reproduction

Directions: *The class will be divided into three major groups, each of which will be assigned a topic. In each group, some students may be the teacher, while others may be students. In the process of discussion, please observe the principles of cooperation, politeness and choice of words. One of the groups will be chosen to demonstrate the discussion to the class.*

1) In the on-site training process, what is the most important focus you should pay attention to?

2) Summarize the rules of on-site training.

3) Suppose you were an engineer of a company, and you should train your workers to form the structure of Dougong. Please schedule the steps to build a Dougong.

Task Four   Discussion and Debate

Directions: *The class will be divided into two groups. Please choose your stand in regard to the following controversy and support your opinions with scientific evidences. Please refer to the specialized terms and classical sentences in the previous parts of this unit.*

In the traditional Chinese culture, ancient Chinese people used wood to build the main frame of their houses. Thus the history of Chinese ancient buildings is known to be "wood history".

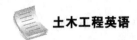

While the people in the Europe in middle ages used stone material. Can you give a reason for this difference?

V. After-class Exercises

1. *Match the English words in Column A with the Chinese meaning in Column B.*

A	B
1) local excavation	A) 大面积开挖
2) mass (bulk) excavation	B) 截水槽
3) soil excavation	C) 少量装药
4) cutoff trench	D) 削坡
5) pump sump	E) 中心掏槽
6) center drift	F) 碾压
7) pilot tunnel	G) 局部开挖
8) light charge	H) 导洞
9) compact	I) 水泵坑
10) cut the slope	J) 土方开挖

2. *Fill in the following blanks with the words or phrases in the word bank. Change the forms if it's necessary.*

formal schedule	new information	expense
substantial improvement	excavation	allocation
estimate	accident	influence
alert		

1) For large projects with many parties involved, the use of _____ is indispensable.

2) Information representation is an area which can result in _____.

3) Updating a project network in response to _____ about a process such as concrete pours can be tedious.

4) _____ problems may easily lead to much lower than anticipated productivity on activities requiring digging.

5) Numerous activities might involve _____ associated with particular cost accounts.

6) The basis of cost _____ would typically be the level of effort or resource required by the different activities.

7) With these allocations, cost _____ for particular work activities can be obtained.

8) Good project managers try to ensure that the job is done right the first time and that no major _____ occur on the project.

9) Safety during the construction project is also _____ in large part by decisions made during the planning and design process.

10) Workers should be constantly _____ to the possibilities of accidents and avoid taken unnecessary risks.

3. *Translate the following sentences into English.*

1）许多商业软件程序可以完成这些试验。

　　_____

　　_____

2）实际项目中时间和成本之间的关系十分复杂。

　　_____

　　_____

3）意外事件可能导致竣工的时间和成本增加。

　　_____

　　_____

4）施工设施中的缺陷或失败可以带来非常大的成本。

　　_____

　　_____

5）成本的增加和工期的拖延就是后果。

　　_____

　　_____

4. *Please write an essay of about* 120 *words on the topic*：*The Beauty of The Traditional Chinese building structure. Some specific examples will be highly appreciated and watch out the spelling of some specialized terms you have learned in this unit.*

_____

_____

_____

_____

_____

_____

_____

_____

_____

_____

_____

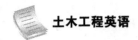
VI. Additional Reading

Brief Introduction on The Forbidden City

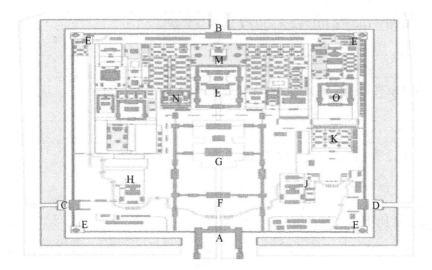

A. Meridian Gate	H. Hall of Military Eminence
B. Gate of Divine Might	J. Hall of Literary Glory
C. West Glorious Gate	K. Southern Three Places
D. East Glorious Gate	L. Palace of Heavenly Purity
E. Corner towers	M. Imperial garden
F. Gate of Supreme Harmony	N. Hall of Mental Cultivation
G. Hall of Supreme Harmony	O. Palace of Tranquil Longevity

[A] The Forbidden City is a rectangle(长方形), with 961 meters (3,153 ft.) from north to south and 753 meters (2,470 ft.) from east to west. It consists of 980 surviving buildings with 8,886 bays of rooms. A common myth states that there are 9,999 rooms including antechambers (前厅), based on oral tradition, and it is not supported by survey evidence. The Forbidden City was designed to be the center of the ancient, walled city of Beijing. It is enclosed in a larger, walled area called the Imperial City. The Imperial City is, in turn, enclosed by the Inner City; to its south lies the Outer City.

[B] The Forbidden City remains important in the civic scheme of Beijing. The central north-south axis(轴) remains the central axis of Beijing. This axis extends to the south through Tiananmen gate to Tiananmen Square, the ceremonial center of the People's Republic of China, and on to Yongdingmen. To the north, it extends through Jingshan Hill to the Bell and Drum Towers. This axis is not exactly aligned(成一条直线) north-south, but is tilted by slightly more than two degrees. Researchers now believe that the axis was designed in the Yuan dynasty to be aligned with Xanadu, the other capital of their empire.

Walls and gates

[C]The Forbidden City is surrounded by a 7.9 meters (26 ft.) high city wall and 6 meters (20 ft.) deep by 52 meters (171 ft.) wide moat(护城河). The walls are 8.62 meters (28.3 ft.) wide at the base, tapering to 6.66 meters (21.9 ft.) at the top. These walls served as both defensive walls and retaining walls for the palace. They were constructed with a rammed(夯实的) earth core, and surfaced with three layers of specially baked bricks on both sides, with the interstices filled with mortar(灰泥).

[D]At the four corners of the wall sit towers (E) with intricate roofs boasting 72 ridges, reproducing the Pavilion of Prince Teng and the Yellow Crane Pavilion as they appeared in Song dynasty paintings. These towers are the most visible parts of the palace to commoners outside the walls, and much folklore is attached to them. According to one legend, artisans(工匠) could not put a corner tower back together after it was dismantled for renovations in the early Qing dynasty, and it was only rebuilt after the intervention of carpenter-immortal Lu Ban.

[E]The wall is pierced by a gate on each side. At the southern end is the main Meridian Gate (A). To the north is the Gate of Divine Might (B), which faces Jingshan Park. The east and west gates are called the "East Glorious Gate" (D) and "West Glorious Gate" (C). All gates in the Forbidden City are decorated with a nine-by-nine array of golden door nails, except for the East Glorious Gate, which has only eight rows.

[F]The Meridian Gate has two protruding(突出) wings forming three sides of a square (Wumen, or Meridian Gate, Square) before it. The gate has five gateways. The central gateway is part of the Imperial Way, a stone flagged path that forms the central axis of the Forbidden City and the ancient city of Beijing itself, and leads all the way from the Gate of China in the south to Jingshan in the north. Only the Emperor may walk or ride on the Imperial Way, except for the Empress on the occasion of her wedding, and successful students after the Imperial Examination.

Outer Court or the Southern Section

Traditionally, the Forbidden City is divided into two parts. The Outer Court (外朝) or Front Court (前朝) includes the southern sections, and was used for ceremonial purposes. The Inner Court (内廷) or Back Palace (后宫) includes the northern sections, and was the residence of the Emperor and his family, and was used for day-to-day affairs of state. (The approximate dividing line shown as red dash in the plan above.) Generally, the Forbidden City has three vertical axes. The most important buildings are situated on the central north-south axis.

Entering from the Meridian Gate, one encounters a large

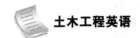

square, pierced by the meandering Inner Golden Water River, which is crossed by five bridges. Beyond the square stands the Gate of Supreme Harmony (F). Behind that is the Hall of Supreme Harmony Square. A three-tiered white marble terrace rises from this square. Three halls stand on top of this terrace(平台), the focus of the palace complex. From the south, these are the Hall of Supreme Harmony (太和殿), the Hall of Central Harmony (中和殿), and the Hall of Preserving Harmony (保和殿).

[G]The Hall of Supreme Harmony (G) is the largest, and rises some 30 meters (98 ft.) above the level of the surrounding square. It is the ceremonial center of imperial power, and the largest surviving wooden structure in China. It is nine bays wide and five bays deep, the numbers 9 and 5 being symbolically connected to the majesty of the Emperor. Set into the ceiling at the center of the hall is an intricate caisson(沉井) decorated with a coiled(缠绕的) dragon, from the mouth of which issues a chandelier-like set of metal balls, called the "Xuanyuan Mirror". In the Ming dynasty, the Emperor held court here to discuss affairs of state. During the Qing dynasty, as Emperors held court far more frequently, a less ceremonious location was used instead, and the Hall of Supreme Harmony was only used for ceremonial purposes, such as coronations, investitures, and imperial weddings.

[H]The Hall of Central Harmony is a smaller, square hall, used by the Emperor to prepare and rest before and during ceremonies. Behind it, the Hall of Preserving Harmony, was used for rehearsing ceremonies, and was also the site of the final stage of the Imperial examination. All three halls feature imperial thrones, the largest and most elaborate one being that in the Hall of Supreme Harmony.

[I]At the center of the ramps leading up to the terraces from the northern and southern sides are ceremonial ramps(土地斜坡), part of the Imperial Way, featuring elaborate and symbolic bas-relief carvings(浮花雕饰). The northern ramp, behind the Hall of Preserving Harmony, is carved from a single piece of stone 16.57 meters (54.4 ft.) long, 3.07 meters (10.1 ft.) wide, and 1.7 meters (5.6 ft.) thick. It weighs some 200 tons and is the largest such carving in China. The southern ramp, in front of the Hall of Supreme Harmony, is even longer, but is made from two stone slabs joined together—the joint was ingeniously hidden using overlapping bas-relief carvings, and was only discovered when weathering widened the gap in the 20th century.

[J]In the south west and south east of the Outer Court are the halls of Military Eminence (H) and Literary Glory (J). The former was used at various times for the Emperor to receive ministers and hold court, and later housed the Palace's own printing house. The latter was used for ceremonial lectures by highly regarded Confucian scholars, and later became the office of the Grand Secretariat. A copy of the Sikh Quanshu was stored there. To the north-east are the Southern Three Places (南三所) (K), which was the residence of the Crown Prince.

Inner Court or the Northern Section

[K]The Inner Court is separated from the Outer Court by an oblong(长方形) courtyard lying orthogonal(垂直的) to the City's main axis. It was the home of the Emperor and his family. In the Qing dynasty, the Emperor lived and worked almost exclusively in the Inner Court, with the Outer Court used only for ceremonial purposes.

At the center of the Inner Court is another set of three halls (L). From the south, these are:

- Palace of Heavenly Purity
- Hall of Union
- Palace of Earthly Tranquility

[I] Smaller than the Outer Court halls, the three halls of the Inner Court were the official residences of the Emperor and the Empress. The Emperor, representing Yang and the Heavens, would occupy the Palace of Heavenly Purity. The Empress, representing Yin and the Earth, would occupy the Palace of Earthly Tranquility. In between them was the Hall of Union, where the Yin and Yang mixed to produce harmony.

[M] The Palace of Heavenly Purity is a double-eaved building, and set on a single-level white marble platform. It is connected to the Gate of Heavenly Purity to its south by a raised walkway. In the Ming dynasty, it was the residence of the Emperor. However, beginning from the Yongzheng Emperor of the Qing dynasty, the Emperor lived instead at the smaller Hall of Mental Cultivation (N) to the west, out of respect to the memory of the Kangxi Emperor. The Palace of Heavenly Purity then became the Emperor's audience hall. A caisson is set into the roof, featuring a coiled dragon. Above the throne hangs a tablet reading "Justice and Honor" (Chinese: 正大光明).

[N] The Palace of Earthly Tranquility (坤宁宫) is a double-eaved building, 9 bays wide and 3 bays deep. In the Ming dynasty, it was the residence of the Empress. In the Qing dynasty, large portions of the Palace were converted for Shamanist worship by the new Manchu rulers. From the reign of the Yongzheng Emperor, the Empress moved out of the Palace. However, two rooms in the Palace of Earthly Harmony were retained for use on the Emperor's wedding night. Between these two palaces is the Hall of Union, which is square in shape with a pyramidal roof. Stored here are the 25 Imperial Seals of the Qing dynasty, as well as other ceremonial items. Behind these three halls lies the Imperial Garden (M). Relatively small, and compact in design, the garden nevertheless contains several elaborate landscaping features. To the north of the garden is the Gate of Divine Might.

[O] Directly to the west is the Hall of Mental Cultivation (N). Originally a minor palace, this became the de facto residence and office of the Emperor starting from Yongzheng. In the last decades of the Qing dynasty, empresses dowager(贵妇), including Cixi, held court from the eastern partition of the hall. Located around the Hall of Mental Cultivation are the offices of the Grand Counciland other key government bodies.

[P] The north-eastern section of the Inner Court is taken up by the Palace of Tranquil Longevity (宁寿宫) (O), a complex built by the Qianlong Emperor in anticipation of his retirement. It mirrors the set-up of the Forbidden City proper and features an "outer court", an

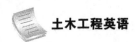

"inner court", and gardens and temples. The entrance to the Palace of Tranquil Longevity is marked by a glazed-tile Nine Dragons Screen. This section of the Forbidden City is being restored in a partnership between the Palace Museum and the World Monuments Fund, a long-term project expected to finish in 2017.

Western Six Palaces
- Palace of Eternal Longevity (永寿宫)
- Hall of the Supreme Principle (太极殿)
- Palace of Eternal Spring (长春宫)
- Palace of Earthly Honor (翊坤宫)
- Palace of Gathering Elegance (储秀宫)
- Palace of Universal Happiness (咸福宫)

Eastern Six Palaces
- Palace of Great Benevolence (景仁宫)
- Palace of Heavenly Grace (承乾宫)
- Palace of Accumulated Purity (钟粹宫)
- Palace of Prolonged Happiness (延禧宫)
- Palace of Great Brilliance (景阳宫)
- Palace of Eternal Harmony (永和宫)

(*If you want to find more information about this topic, please log onhttps：//en. wikipedia. org/wiki/Forbidden_City*)

1. *Read the passage with ten sentences attached to it. Each statement contains information given in one of the paragraphs. Identify the paragraph from which the information is derived. You may choose a paragraph more than once. Each paragraph is marked with a letter.*

   1) The Palace of Earthly Tranquility (坤宁宫) is a double-eaved building, 9 bays wide and 3 bays deep.

   2) The Hall of Central Harmony is a smaller, square hall, used by the Emperor to prepare and rest before and during ceremonies.

   3) The Forbidden City remains important in the civic scheme of Beijing. The central north-south axis remains the central axis of Beijing.

   4) It mirrors the set-up of the Forbidden City proper and features an "outer court", an "inner court", and gardens and temples.

   5) They were constructed with a rammed earth core, and surfaced with three layers of specially baked bricks on both sides, with the interstices filled with mortar.

   6) In the south west and south east of the Outer Court are the halls of Military Eminence (H) and Literary Glory (J).

   7) These towers are the most visible parts of the palace to commoners outside the walls,

and much folklore is attached to them.

8) The Hall of Supreme Harmony (G) is the largest, and rises some 30 meters (98 ft.) above the level of the surrounding square.

9) It was the home of the Emperor and his family.

10) Only the Emperor may walk or ride on the Imperial Way, except for the Empress on the occasion of her wedding and successful students after the Imperial Examination.

1) _____ 2) _____ 3) _____ 4) _____ 5) _____

6) _____ 7) _____ 8) _____ 9) _____ 10) _____

2. *In this part, the students are required to make an oral presentation on either of the following topics.*

1) The Construction style of the Forbidden city.

2) What part of the Forbidden City impressed you most?

习题答案